The Art of War

Christian Feest

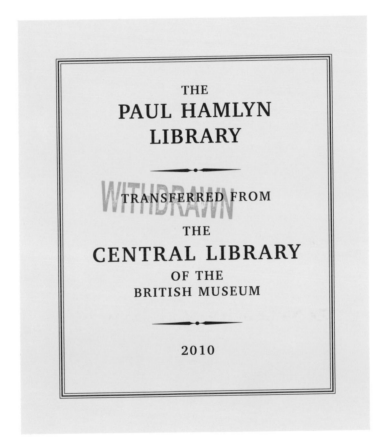

THAMES AND HUDSON

First published in the UK in 1980 by
Thames and Hudson Ltd, London

© John Calmann & Cooper Ltd, 1980
This book was designed and produced by
John Calmann & Cooper Ltd, London

Filmset by Southern Positives and Negatives (SPAN), Lingfield, Surrey.
Printed in Hong Kong by Mandarin Offset International Ltd

Contents

Introduction

Wherever people live together, conflict is bound to occur. War results when force is organized to settle a conflict between two groups of people. This use of force is sanctioned by the community and always follows certain rules of conduct, which are not the same as those obtaining in times of peace. In particular, killing is generally regarded as legitimate in war.

The phrase 'art of war' has several meanings: it can apply to military strategy, to skill in the handling of offensive and defensive weapons, or to the aesthetic aspects of warfare. The third meaning especially deserves further comment. Both 'art' and 'war' are Western concepts for which it is often impossible to find exact equivalents in tribal philosophy. An object that appeals to our aesthetic sense probably appealed to its maker because of its meaning and usefulness. Just as the artifacts of the tribal world form an integral part of the lives of their makers, and cannot properly be considered separately, so, too, the form that war takes is a product of the society that is waging it. While it is not possible in this book to discuss all the various concepts of war held by tribal peoples, it should be clear that 'war' does not mean the same thing everywhere: it could hardly do so given the range of economic and social diversity among tribal man. Among hunters and gatherers, for example, both the conflicts and the means of resolving them differ from those of pastoralists or irrigation farmers. But it would be simplistic to try to explain all the differences on the basis of subsistence economy alone.

A recurring factor among the motivations for waging war is the competition for scarce resources in the broadest sense. The variety of objectives for which people are willing to risk their lives can be governed by ecological and demographic factors, the prevalent type of subsistence economy, or the social organization of the warring tribes. Disputes over the use of territory or access to economic goods and resources are by no means universal reasons for warfare. The Jivaro of Ecuador and Peru, like many other tribes, would not consider embarking on territorial conquests because they fear and detest the country of their enemies. Objectively they have no need to expand their territory because there are no relevant economic or population pressures on them. In East Africa, on the other hand, quarrels over grazing rights are the main cause of armed strife: herds are the principal form of wealth, and the only way to increase wealth is to gain access to more grazing lands.

In a sociological sense, the scarce commodities that can be acquired by warfare are status and prestige. In loosely structured societies, success on the warpath is often the key to leadership. And even where rank is generally ascribed to individuals on the basis of inheritance, achievements in war can definitely improve a person's

1. Unlike soldiers of modern armies, the warriors of many tribal societies dress up for battles, wearing beautiful clothes and ornaments or painting themselves. The weapons of this Dyak portrayed in 1879 consist of a sword and blowgun lance. His armour includes a rattan helmet, a fur breast plate and a wooden shield. *Lithograph from C. Bock,* **The Head Hunters of Borneo,** *1882*

social standing. In some societies, to have many wives is regarded as indicative of status: the normal sex ratio prevents polygyny from becoming the usual form of marriage. The relative scarcity of women is therefore frequently the cause of conflict, at least in part. While male enemies are killed in combat or after having been taken prisoner, women are spared. The correlation between warfare and polygyny (which is by no means universal) is increased by the number of male casualties in combat, which also distorts the usual numerical relationship between the sexes and improves the chances of polygynous marriage.

Wars are also frequently fought to protect the rights, property and lives of members of the tribe or group. Such wars can be either defensive or, as in cases of suspected witchcraft on the part of the enemy, a pre-emptive measure, offensive. Revenge, at first sight, generally seems to be only a minor motive, but tends to become the primary reason if there is no other way of resolving the conflict. The question of who originally started the conflict is of little importance. All that matters is that the latest attack has to be countered.

An almost universal device employed to transcend the natural barrier against killing is to assume that an enemy group is inferior to the point of belonging to a different species – thus, it is fairly common to designate enemy tribes as 'snakes' or other kinds of dangerous vermin. Accusations of cannibalism, witchcraft, or other crimes against humanity levelled against the opponents provide a simple rationalization for denying their status as human beings. In complex societies, the wrong religion often identifies the 'heathen' or 'non-believer' as an object of justified annihilation. These attitudes are reflected in the way wars are fought. In fights between subdivisions of the Nuer of Sudan no houses are destroyed, nor are women and children killed, but no such restrictions apply in armed conflicts with outsiders.

Even if the strangers are belittled as somewhat lacking in terms of human status, the suspicion often remains that they possess powers beyond the control of the 'home' group. Warfare can thus serve to demonstrate superiority over others, and at the same time to increase the warrior's store of supernatural power, which he takes from the enemy he has killed. This leads to the point where some tribes regard warfare as necessary for maintaining cosmic harmony. Head-hunting is a classic example of this: since supernatural power is the scarce commodity sought and an important way of acquiring it is by taking the heads of enemies, head-hunting expeditions are essential.

Some, but not all, wars replace litigation in the absence of agreement about legal procedure in peacetime. Raiding parties may approach enemy territory in an attempt to obtain what they believe is

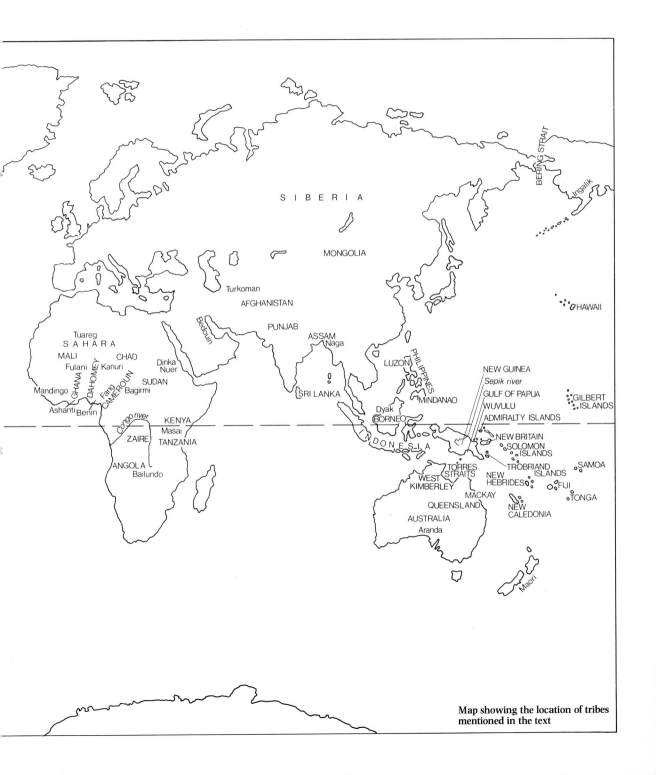

Map showing the location of tribes mentioned in the text

2. *(Left)* **Occasionally tribal warriors depicted themselves killing Western soldiers, as in this drawing by a Sioux warrior of himself shooting a U.S. soldier. From the autobiography of Tall Bear, 1874.** *London, British Museum*

4. *(Right)* **An early Port Jackson watercolour showing a European fleeing from an Australian aborigine, who is about to launch a spear with a throwing stick. The resistance of tribal peoples to colonial exploitation led to a common misconception that they were 'warlike savages'.** *London, British Museum.*

3. *(Below)* **A monumental painting on bison skin probably made by a Pueblo Indian. It shows the massacre of the Spanish Villasur expedition of 1720 by Pawnee and Oto Indians in Nebraska. None of the Indians are using firearms, but bows and arrows and lances are supplemented by European-made iron hatchets and swords.** *Lucerne, Collection Hans Ulrich von Segesser*

6. *(Above)* A group of Turkoman warriors attacks a Russian exploring party armed with Persian sabres and short daggers held between the teeth and using inflated sheep-skin floats. Against regular troops such traditional tactics proved to be of limited value.

5. *(Left)* Native military units possessing the advantages of a detailed knowledge of local conditions were established by all colonial powers. In such auxiliary troops, tribal people could continue to pursue some of their traditional martial values. The mixed native and military garb of the Bontoc Igorot constabulary soldiers reflects their intermediate position between opposing tribal and Western forces.

rightfully theirs. Battles may be fought to decide an issue after negotiations have failed. But, at least in those cases in which the motivation is not religious, there may be other possibilities of resolving conflicts than by the use of force.

Human experience has shown that the art of peace is more difficult to master than the art of war. The formal peace-making which follows a state of war will be discussed later; apart from that there are peaceful substitutes for armed conflict. The kind of tension which could lead to organized violence is frequently relieved by competitive games or other means of diverting serious harm from society. The Eskimo have become noted for the fact that they do not wage wars amongst themselves. Singing duels, in which the opponents are publicly ridiculed, are held to resolve disagreements or to reduce any tensions within the society. An explanation of this peaceful habit is that healthy males are the most precious resource of a particular group. Because women can contribute little to the subsistence of the tribe, which depends largely on sea-mammal hunting, some Eskimo bands even used to practise selective female infanticide. It would seem incongruous to risk male lives needlessly in armed warfare.

The 'potlatches' – feasts during which individuals and groups competed in the display, distribution, and destruction of property to raise their prestige – were regarded by the Kwakiutl of British Columbia as a substitute for war. The flourishing trade in sea-otter

7. A painting by George Catlin of 'Little war', one of the native names of a ball game played by many tribes in the southeastern United States. Members of different towns or sometimes tribes formed the opposing teams. Players were subject to rituals and taboos before and after the game which were similar to those imposed on warriors going into or returning from war. The game (which is still played by some tribes) is very rough, serious injuries being not uncommon. *Oil on canvas, 19½ × 27½in (49·5 × 70cm). Washington, Smithsonian Institution*

8. Intertribal wrestling matches are an important means to channel latent hostilities between tribes of the Xingú area of Brazil. Contestants prepare themselves carefully for the event; they follow strict rules during the fight and exchange embraces afterwards

furs which had sprung up after European contact had brought peace to Kwakiutl society. This trade brought instant and enormous economic advantages for the tribe, as well as giving the Hudson Bay Company, the major trader, a considerable influence over their way of life. The traders were anxious to keep intertribal strife to a minimum because it interfered with their economic interests. More important, however, was the fact that the affluence associated with trade offered the Kwakiutl some of the benefits that had previously been sought in warfare. Other goals of war, such as improved status, were now channelled into potlatching.

This example is not quite typical of the relationship between trade and warfare in general. As in the Kwakiutl case, trade may diminish the importance of warfare, offering a less dangerous way to get access to scarce resources. But once the goods offered in exchange them-selves became less than plentiful, competition for them will increase the chances of armed conflict. Many of the Indian wars of the early colonial period in eastern North America were indeed fought to gain better hunting and trapping territories.

War and Peace in Tribal Societies

Military Organization

The organization of military affairs is usually separate from the political organization which exists in times of peace. Warfare necessitates a type of behaviour – killing, looting, and other acts of aggression against persons and property – that most societies would regard as highly unacceptable and which would under ordinary circumstances demand punishment rather than praise. No war can be fought without breaking the codes of civilian life. Depending on their relative needs for protection and aggression, different societies have devised different ways of solving this problem.

The distinction between condoned and condemned violence is effected by separating the roles of warrior and peaceful citizen. Standing armies and professional soldiers are, however, rare in tribal societies. Usually all the members are called upon to contribute to the defensive and offensive requirements of the group. Everybody able to bear arms is expected to support the common cause. What varies in times of war and peace is not so much the personnel but its organization. The following examples show varying degrees of distinction between military and civil organization; the extent of this separation is not necessarily dependent on the complexity of the society in general. The Jivaro ordinarily do not have a chief because there is no need for a centralized political authority. Day-to-day decisions are made by the heads of kinship groups; they also settle disputes arising between those groups. Only in times of war will the Jivaro submit to the authority of an experienced warrior, whom they call 'the old one' and whose commands in matters of war they obey strictly. Here, as in many similar cases, a war-chief is selected on the basis of proven ability, which illustrates the general rule that military leadership is more frequently based on achievement than is civil authority. The size of a raiding party will usually depend upon the reputation of its leader, whose following will increase with success and diminish with failure.

A more formalized distinction between the organization of a group in war and in peace is illustrated by the case of the Cherokee, a farming tribe of the south-eastern United States. Each Cherokee town, a politically autonomous unit, was organized by two complementary authorities. White or Peace chiefs, consisting of clan leaders and respected elders, would rule by consensus in all matters relating to the internal affairs of the community. Red or War chiefs, drawn from the ranks of younger warriors, were responsible for the town's external relations, including trade, hunting expeditions, and warfare; in their sphere they supplied assertive leadership. All townsmen were members of both organizations. Which of the two took precedence depended on the situation prevailing.

9. Photographic records of tribal warfare are rare. This one shows the frontlines of two inimical Dani groups of the New Guinea highlands cautiously approaching one another during one of the scheduled one-day battles which are fought regularly between neighbouring groups.

11. Almost as soon as they learn to walk, Jalé boys from the central highlands of New Guinea playfully compete in shooting reed stalk arrows at moving targets. This early training in the use of bows and arrows is essential for boys who will grow up to be hunters and warriors.

Frequently there is a strong association between hunting and war wherever hunting is the primary male economic activity. Not only are some of the weapons used for both purposes, but the principles of male cooperation are to be found in both communal hunts and the organization of war parties. The inclusion of hunting in the domain of the Cherokee's Red organization is an illustration of this association.

By contrast, male members of the Dinka of Sudan either belong to a hereditary class who have a monopoly of ritual power and are exempt from military service, or else they are warriors. The religious specialists contribute to military affairs by praying for victory. There is a tendency to select war leaders from a specific kin group, but in practice ability counts for more than inheritance.

An example of even more specialized organization is supplied by the military societies of the Nothern American Plains Indians. These fraternities provided bonds across kinship groups, policed important communal events (such as the bison hunt) and acted as military units in times of war. Membership of such a society was indicated by special regalia, by which officers could also be distinguished from rank and file members. Within the fraternities a warlike spirit was cultivated. Membership sometimes entailed the obligation to be brave up to a certain point, for example, not to retreat behind a lance rammed into the ground. While some of these societies were age-graded, so that throughout his life a man would acquire membership in all of them in

10. Kikuyu shield used in dances accompanying circumcision. *17½in (45cm). London, British Museum*

succession, others were not hierarchically ordered and members competed to outdo one another.

Among the Masai of Kenya and their pastoralist neighbours, on the other hand, the class of warriors is defined strictly on age-grade principles. Those who, by a ceremony of circumcision, have passed the stage of boyhood, join the ranks of the warriors until they get married, at which time they retire from active military duty. The warriors live segregated from the rest of the community and form a strongly integrated group. They are not allowed to drink beer, but are entitled to promiscuous sexual relations with unmarried girls.

Only with the emergence of a strong central authority do professional bodies of warriors gain prominence. In African kingdoms (but also in some other centralized chiefdoms) there were usually certain elite troops, such as border or palace guards, which were specially trained and supported by tributes paid to the ruler. Their duties might include construction and repair of public buildings to keep them busy during times of peace.

In tribal societies vocational specialisation is generally of lesser importance than the division of labour by sex. The limitations imposed on women by their roles as mothers mean that activities which involve prolonged absence from the household tend to fall to the men. Hunting and warfare thus become typically male domains. There are secondary motivations for the exclusion of women from military assignments, such as the widespread belief that menstrual blood or other manifestations of femininity may weaken or destroy the male powers necessary to succeed in war. Women are thus frequently prohibited from touching weapons (many of which carry phallic connotations), or coming into close contact with men preparing to go into battle. The Apache warriors of the American Southwest carefully avoided pregnant women, perhaps because their condition symbolized the power to give life and might diminish the men's power to dispense death.

Nevertheless, there are exceptions to the rule. While the Amazon society of classical tradition, in which bellicose women reverse the usual role, corresponds to no ethnographic reality, women sometimes could become warriors. Just as some Plains Indian men who felt they were not up to the physical and psychic demands of warfare might decide to become transvestites, specializing in female skills, strong-hearted women could take up arms, obtain war honours, and achieve fame as warriors. But even they were not usually allowed to display the insignia corresponding to their martial achievements.

In most societies, however, women did not participate in the actual fighting but would support their husbands in other ways. Jivaro women danced at home to enlist supernatural support for their men

12. In the highlands of New Guinea, an old man teaches his grandson how to use a bow and arrow.

13. A left-handed and a right-handed Ahaggar-Tuareg engage in a fighting match with sword and shield, surrounded by the stark landscape of the Sahara desert. The shields, made of the skin of a gazelle, are very light and strong, although they can be split by a sharp sword.

in enemy territory; those women of the Ojibwa of Wisconsin who accompanied the warriors on military ventures were permitted to mutilate the dead by cutting off their genitals. Elsewhere, they played the part of camp followers, encouraging and feeding their warrior husbands. The unmarried women of the Arab Bedouins stimulated the warriors by threatening to offer themselves to the enemy, exposing their breasts, unveiling their faces and undoing their braided hair.

Preparations for the warpath

Since in most tribes all boys were eventually to become warriors, it was important that they acquired the essential skills. Their childhood play would emulate adult life: war games were organized to teach them tactics and develop qualities of leadership, while they practised with toy weapons. Stealing upon an object representing the enemy was as important to an Apache boy as being able to run with a mouthful of water without spilling a single drop, or slinging and dodging stones. In a critical situation, all this could save a man's life.

Similar war exercises continued in adult life, although they varied in frequency and elaboration from tribe to tribe. In some militaristic societies regular mock battles were staged: the Fulbe and other Sudanese tribes held knightly tournaments for mounted warriors, while the Tuareg of the Sahara held sham fights with sword and shield.

Since standing armies are not generally kept, recruiting has to take place, at least for offensive warfare. When it is under attack, every member of the community will voluntarily protect it as well as himself. But for raids or other acts of aggression, potential warriors will have to be given some incentive. It may be sufficient for the chief or whoever is in charge of the military operations to invite the warriors to join the party. On the Trobriand Islands off the coast of New Guinea, the warriors retire at this point to the capital of their district where they remain until the end of the war. In most cases, however, more active recruiting is needed. In the Asmat area of New Guinea large wooden poles representing the ancestors are erected, to remind everybody that they died at the hands of the enemy. The ceremony deeply stirs the emotions of the men, who believe that their ancestors continue to exert a vital influence on the world of the living. The raising of the pole puts them under a severe moral obligation to avenge the particular ancestors represented.

Before going on the warpath every warrior does his utmost to ensure a successful outcome. In almost all cases there are restrictions on the behaviour of the warrior and his family before, during or after contact with the enemy. Most of these are aimed at conserving his

14. A stylized version of a duel is performed by two Buru warriors as part of a war dance for the Dutch resident of their island. The narrow wooden shields of the Moluccas, decorated with shell inlays, are typical parrying shields especially useful in close combat with striking weapons. *Lithograph from J. Dumont Durville,* **Voyage de la Corvette l'Astrolabe,** *1833.*

physical and spiritual powers or at protection against the powers of his opponents. These taboos can range from the prohibition of sexual intercourse to forbidding scratching the head with the fingernails. The Apache even used a special warpath language which avoided certain words.

At the same time it is of crucial importance to rally all the supernatural help available. Many North American tribes used war bundles, the contents of which reflected instructions received from a spirit by the owner in a vision. In times of need, the owner would open the bundle and perform the rituals required in order to obtain the assistance pledged by his supernatural benefactor.

Omens portending the outcome of the battle are looked for and interpreted by specialists. The Dani of the central highlands of New Guinea tie up grasshoppers or small rodents with grass, to represent the enemies that will die in the approaching fight. The Murngin of Arnhemland throw their spears at two specially made ceremonial spears, one of which represents their own group and the other the enemy. The hits and misses scored during these preliminaries predict the outcome of the struggle.

Among the Trobriand Islanders a ritual specialist is employed to chant over the shields to make them spearproof, to treat the warriors to make them strong and enduring, and even to cast a spell on the battlefield to make the enemy run away. War dances held before the fight serve as ritualized instructions on how to behave in times of crisis

Dessiné d'ap. nat. par le voyageur Douville et lith. par M'. B.   Lith. de Engelmann.

Soldat Bailundo partant pour la guerre.

(Sa première femme et ses enfants qui l'avaient accompagné retournent au village.)

Publié par Jules Renouard, libraire à Paris. 1832.

16. An image of Kuka'ilimoku wearing a crested helmet and representing the Hawaiian god Ku as war god of the Kamehameha dynasty. Figures such as these were made of a trellis covered with a net in which precious red feathers were fastened. The eyes are made of mother of pearl, while the mouth is set with dogs' teeth. These images, to which human sacrifices were made, were either kept in temples or carried into battle on long poles. *Honolulu, Bernice P. Bishop Museum*

17. In mythic times, the twin war gods of the Zuni of New Mexico instituted the Priesthood of the Bow and instructed the priests in the secret and magic rituals necessary for successful warfare. During an annual winter ceremony their images, preferably carved from lightning-struck pine wood, were used and afterwards deposited together with prayer sticks in shrines near the mountainous abode of the gods. *30½in (77·4cm). London, British Museum*

15. A Bailundo warrior of Angola leaves for war while his wife and children, who have accompanied him on part of his way, return to their village. His only weapon is a flintlock gun which, by the early nineteenth century, had replaced all native arms. *Coloured lithograph from J. B. Douville, **Atlas du Voyage au Congo,** 1832*

and are in a sense 'previews' of things to come. Sometimes they include mock battles, which are regarded as magical means of ensuring victory. War dances also strengthen the feeling of mutual responsibility and *esprit de corps*.

The dress and ornaments worn by the warrior often included symbols of strength, endurance, valour, audacity, swiftness – in short all the qualities he wished to manifest himself. Being overdressed, however, could impede the freedom of movement, and so many tribes would strip down rather than dress up for battle. The Naga of Assam, for example, donned their regalia only in pitched battles, not on raids. Even almost naked warriors were not completely without means of decoration: body and face painting usually had a specific symbolic meaning. Some North American tribes, like the Salish, painted their faces half red and half black. The red half was to guarantee good luck for the wearer, the black half spelled doom for the enemy. Other decorations were intended to frighten the adversaries: the Jivaro, for instance, covered themselves with black paint to give the impression that they were demons. Scare tactics also included hair-raising war cries.

Strategy

Warfare could be waged by means of pitched battles or raids. Since the overriding object in most kinds of tribal warfare is to avoid casualties while inflicting some damage to the enemy, raids, which have the element of surprise, are generally preferred. This in turn leads to an increase in defensive measures, stemming from the desire not to be caught off-guard. Some societies, like the Trobriand Islanders, would tolerate only previously announced fights, but in this they were an exception.

Usually, small groups of warriors steal into enemy territory to attack an unsuspecting village at dawn, to kill solitary workers in the fields, or to lie in ambush for their victims. Sometimes the raid is aimed at the acquisition of property, such as horses, rather than the taking of human lives. After the successful accomplishment of their mission, the invaders quickly return home before the enemy gets a chance to cut them off. Actual fighting is not necessarily sought but often cannot be avoided.

Raiding is a good strategy for inflicting damage, demoralizing the enemy, and acquiring movable property, war honours, or trophies. It does not resolve a dispute but invites a vicious cycle of attacks and counter-attacks. Pitched battles on the other hand are far more formalized and are governed by rules which protect the safety of non-combatants and of civil life in general. Often they are regarded as adequate means of settling conflicts.

18. After having been officially recognized by the community, prestigious war deeds were recorded in scenic paintings on bison skin by Plains Indian warriors as on this early nineteenth-century Hidatsa robe. Concerted strategy was a less important aspect of warfare on the Plains than individual acts of courage. *118in (300cm). Stuttgart, Linden-Museum*

20. Armed with inlaid wooden clubs, spears and spearthrowers, Tapuya men of northeastern Brazil imitate fighting action in their war dance to heighten their enthusiasm for the forthcoming encounter with the enemy. *1641. Oil painting by Albert Eckhout. Copenhagen, National Museum of Denmark*

21. *(Right)* Drawn by a Jesuit missionary in Paraguay around 1760, this scene of a battle between two tribes of the southern Gran Chaco shows most warriors more or less naked. Only the leaders in the front wear special costumes such as jaguar-skin mantles and bird-skin headdresses. Even though Indian tactics were aimed at avoiding casualties, some encounters were obviously fairly bloody. *Watercolour drawing by Florian Baucke. Collection Zwettl*

19. A wooden ancestor figure placed near the burial platform illustrates the full regalia worn by a Konyak Naga when going into an open battle. *27½in (70cm). Vienna, Museum für Völkerkunde*

Battles are customarily agreed upon between opposing parties after one side has challenged the other. Both groups assemble at the set time and place. After a little rest, a good meal, and last-minute preparations, battle is more or less formally joined. Among the Dani such encounters last from about noon until late afternoon, unless rain or excessive heat exhaust the warriors earlier. Short periods of active fighting alternate with longer intermissions to give the participants a chance to recover from their exertions, exchange verbal insults, and attend to the wounded. The fighters have to move about constantly to avoid being hit by arrows and spears. The fighting ends when the warriors – afraid of the spirits of the night – leave the battlefield to get home before dark.

Among the Naga, pitched battles last from half an hour to two hours. As soon as one side feels unable to withstand the pressure of the enemy's attacks, the warriors turn to flight, covering their retreat by mining the way with calthrops. A limited number of tactical ruses is used, mainly to try to ambush the enemy. Thus, the main force may withdraw as if overpowered, the better to attack from the flank or the rear.

Another widespread form of previously arranged combat takes the form of a number of duels fought simultaneously between members of opposing groups in an effort to settle their differences. The Aborigines of Queensland arm themselves with long heavy wooden clubs and shields; they alternately strike one another and ward off blows until either the shield breaks or one of the combatants tires and gives up. At this point the old women intervene, begging for the loser's life. Casualties are consequently few.

The availability of horses greatly affects strategy in that it enables the warriors to converge swiftly on the target of their attack, while being themselves less vulnerable. The Dakota, and other tribes which had been pushed out on to the North American Plains by tribes possessing firearms, became feared and respected warriors after they had acquired the skills of horsemanship. Nevertheless, many tribes prefer to use their mounts mostly for getting to and from a battle and to do the actual fighting on foot.

Where large expanses of water separated the enemies, boats had to be employed. On the Northwest Coast of North America the condition of the terrain largely limits overland travelling, and attacks therefore were always made from the sea. Coastal villages were approached in huge wooden dugouts whose high wide bows shielded the warriors before and during landing. Some had loopholes through which arrows could be shot. Headhunting raids by the Solomon Islanders were similarly carried out in special war-canoes whose sides were inlaid with mother-of-pearl.

22. Martial music is not limited to Western civilizations. Flutes were used by North American Indians to rally the warriors, while the sound of drums spreads the news of the enemy's approach in many parts of the world. Signals on trumpets like this one from Zaire, supposedly also used as a club, helped to co-ordinate the advance or retreat of troops. *15in (38cm). Vienna, Museum für Völkerkunde*

23. Brandishing a spear and holding a trophy skull encased in a basketry cone, a Naga headhunter performs a cermonial dance. The basket attached to his belt serves to carry wooden calthrops into battle.

24. Hostilities arising out of conflicts over hunting rights or women between different Botocudo kinship groups of eastern Brazil were fought out in a series of duels. The men pounded one another with long poles, while the women engaged in tearing their adversary's hair, lip plugs and ear plugs. *Engraving from Maximilian, Prince of Wied-Neuwied,* **Reisen seiner Durchlaucht des Prinzen Maximilian von Neuwied nach Brasilien**

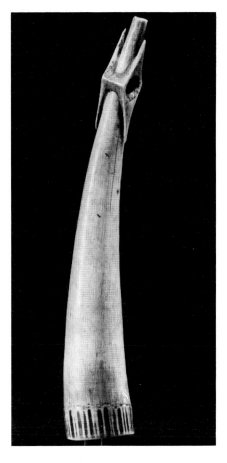

25. *(Left)* **Excellent horsemen, the Comanche Indians of Texas would use the bodies of their horses as shields in close encounters with their enemies by letting themselves drop down on one side of the horse while it was galloping at full speed. To be able to do this safely, use their weapons while covered by their horses, and then return to an upright position, young warriors had to practise a lot in sham battles.** *Oil painting by George Catlin. Washington, Smithsonian Institution*

True naval engagements were not common in tribal warfare, but Polynesia, with its highly developed seacraft, was a major exception. Large numbers of boats, sometimes lashed to one another to prevent the formation from breaking up, were involved in these battles. First slingstones and spears were thrown from a distance, and the fighting continued with hand-weapons as soon as the enemy's boats were rammed and boarded. On a smaller scale were the battles fought at sea by the Melanesians of the Admiralty Islands, whose chiefs first met in single combat before their supporters joined the scuffle, or by the tribes of eastern Canada, who tried to sink their opponents' fragile bark canoes by hurling rocks at them.

A Permanent State of War

If fighting is not aimed at settling a specific conflict, such as a dispute over property, but instead is undertaken primarily for more general goals such as the acquisition of status or supernatural power, or because of the belief that war is necessary for the continual regeneration of the world, a lasting peace is not sought and indeed is not really possible. Few societies, however, can afford to live in a permanent state of war. Intensive fighting has to stop to allow the economy to supply the vital needs of the fighting tribes. Agreement to

26. *(Above right)* **After a successful attack on a Clallam village in the Straits of Juan de Fuca, the Makah return in their splendidly painted war canoes proudly displaying the heads taken in revenge for the murder of two of their men.** *Oil painting by Paul Kane. Toronto, Royal Ontario Museum*

27. **Tahitian war canoes consisted of two boats joined by transverse timbers. Bows and sterns rising up to 21 feet (7 metres) above the water were carved and fitted with variously marked pieces of cloth identifying the naval units. In 1774 James Cook witnessed a review of the fleet consisting of 160 big war canoes and 160 smaller boats, the latter apparently ordinary canoes pressed into war service.** *Oil painting by William Hodges. Greenwich, National Maritime Museum*

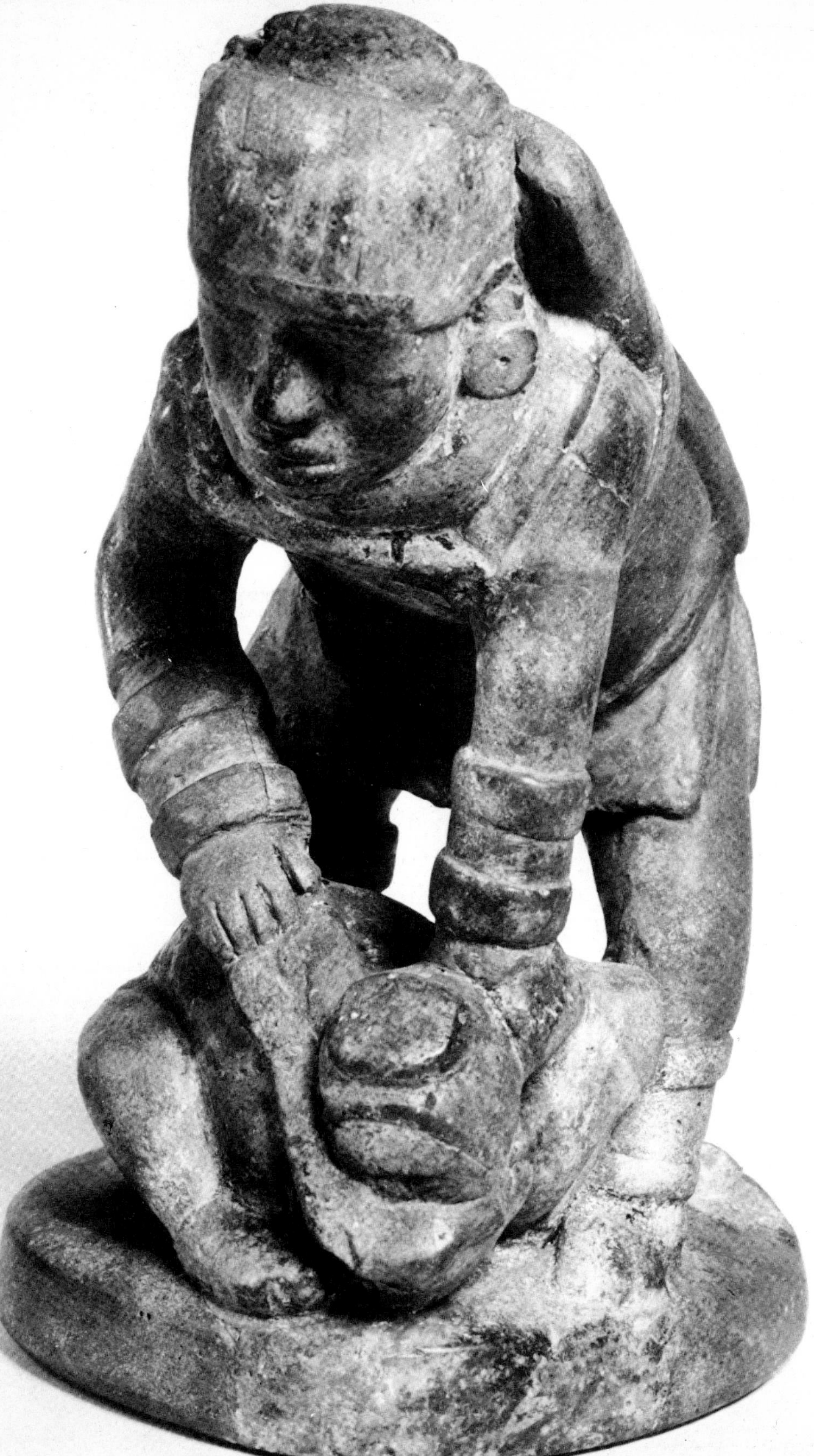

29. *(Above)* **Sixteenth-century Europe was shocked by reports about cannibalistic practices overseas, especially in South America. Illustrations like this one ('How the savages roast their enemies') left no detail untold. Disregarding their own acts against humanity, Europeans used cannibalism as an excuse for enslavement and genocide of native populations.** *From André Thevet,* La Cosmographie Universelle, Tome Seconde, *1575*

28. Leaning over his subdued enemy, a warrior whose back is protected by a piece of armour proceeds to take a head trophy. This effigy pipe, which presents a rare insight into prehistoric North American Indian warfare, was recovered from Spiro Mound, Oklahoma, and dates to the period between AD 1200 and 1600. 9¾in (25cm). *New York, Heye Foundation (Museum of the American Indian)*

a truce may be reached, in which case frequently the long-term result is that fighting becomes a markedly seasonal activity.

Headhunting and cannibalism are sometimes classified as war customs. It is, however, often more meaningful to view them as religious practices which provide the impetus for a permanent state of war. Occasionally, head trophies have become simply tokens of military achievements or a means of gaining prestige, but for many headhunting tribes they are ritual objects charged with supernatural power. The enemy's body, when eaten by the victors, is likewise regarded not as food but as symbolizing a mythic being to be devoured in a ritual re-enactment of mythic events. For their practitioners these are not gory incidents in war: they are the causes for which war is fought.

In a myth widespread in Melanesia, mankind is said to have originated as the result of a struggle between a giant Sky God (who is also the Lord of the Animals) and the fatherless twin children of the Earth Mother. The boar-like Sky God threatens to kill and eat his fellow gods until he is himself killed by the twins. By eating his body the twins gain the power of procreation and sire the ancestors of

mankind. The practice of cannibalism, by imitating the myth, serves to renew these vital powers which bestow fertility not only on man but also on crops and nature in general. A similar ideology seems to lie at the root of cannibalistic rituals in tropical South America.

Headhunting and the torturing of prisoners-of-war to some extent involve similar concepts of power transfer. The Dyak boys of Kalimantan, for example, will be regarded as men and ready for marriage only after they have returned from the warpath with an enemy's head. Where this sort of custom no longer has a religious foundation the original idea has degenerated. The rapid spread of scalping in North America during colonial times illustrates how the custom could be perpetuated even after its religious significance had disappeared. Cannibalism and headhunting do not necessarily indicate valour or other warlike ideals. Since the purpose of the raids was to return with a human head, the Kalinga of Luzon considered themselves lucky if they could ambush a lone man, an old woman, or a child, because it was less dangerous to kill them. For the same reason, scalping was considered only a minor accomplishment by the Plains Indians when compared with being the first to touch the enemy's body, a custom known as 'counting coup'.

Heads brought back from a headhunting expedition could be treated in various ways. Sometimes the skull was only skinned and perhaps adorned, but frequently attempts were made to retain the appearance of the face. In the Sepik area of New Guinea, facial features were restored after the skinning by modelling them with clay and applying paint, while other tribes of the island made less realistic

30. European observers (including writers and illustrators of fiction) were fascinated by the equanimity exhibited by North American Indian prisoners of war when tortured by their captors. For the captive this stoicism amounted to a display of the powers he acquired by contact with the supernatural; for his tormentors it may have increased the value of their victory. *From G. L. Jerrer,* **Neue Bilder-Geographie für die Jugend**, *1819*

31. Head trophies were taken by the Mundurucú to increase the magical power of the victor and to increase fertility of man and nature. The dried heads were sometimes decorated with symbolic reference to the sun and were carried by their takers wherever they went. *Height (excluding rope): 8in (20·5cm). Berlin, Museum für Völkerkunde*

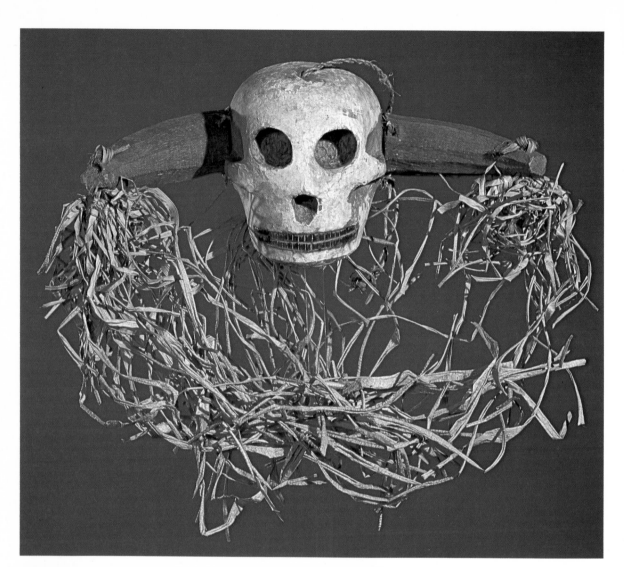

32. If a Naga killed an enemy but was prevented from taking his head, he could use a wooden substitute in its place. Even the mithan horns which customarily embellish the trophy are executed in wood. *16in (40cm). Vienna, Museum für Völkerkunde*

33. *(Left)* Scalping, an abbreviated form of taking a head trophy, is widely regarded as a highly characteristic North American Indian war custom. It was the introduction of steel knives and the payment of bounties for enemy scalps, however, which furthered the diffusion of a custom originally limited to a few tribes in eastern North America. *Oil painting by Peter Rindisbacher. West Point, U.S. Military Academy*

34. *(Right)* The Maori preserved the heads of dead enemy chiefs to insult them; they were impaled on sticks and exposed to public vilification. *Height 8in (20cm). Basle, Museum für Völkerkunde*

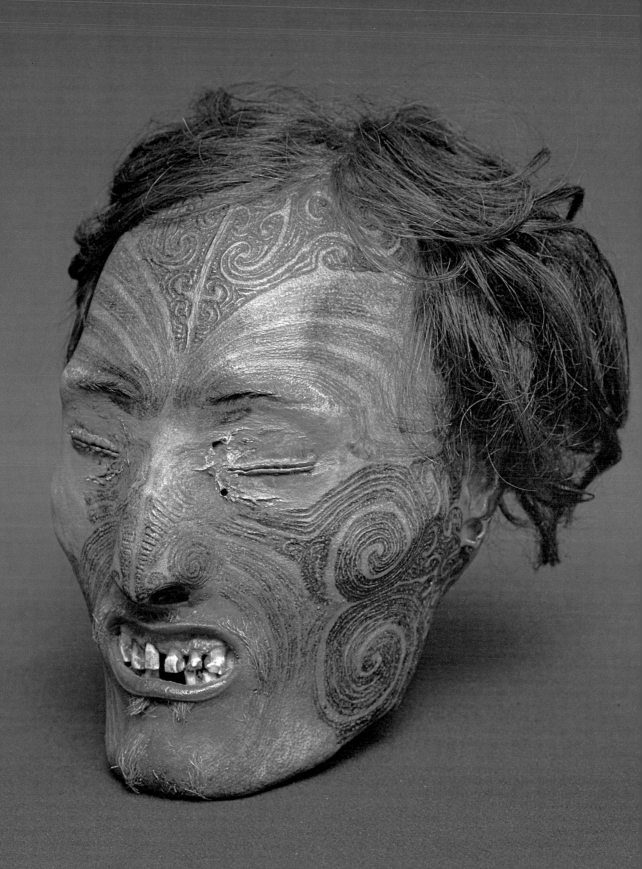

constructions, ornamenting the skull with protruding tubular eyes set with shell beads, or with an enormously elongated nose. The Maori of New Zealand and the Mundurucú of Brazil removed the brains from the skull but dried the skin until it resembled parchment.

The Jivaro used only the skin in the preparation of their *tsantsas*. After a vertical cut had been made in the back of the head, the skin was carefully drawn off the skull and boiled, and the cut sewn together again. Next the skin was repeatedly filled with hot sand and stones and the remaining flesh scraped off from the inside until the head was reduced to about a quarter of its original size. Finally, the trophy was painted black. Due to the enormous and somewhat sick demand for such shrunken heads by a Western market, Jivaro *tsantsas* have become the most widely faked trophies.

Whereas the shrunken heads were prepared from heads brought back from the warpath, scalps could be taken on the spot from a dead or unconscious victim. They could vary in size from small circular pieces to major portions of the skin. Some Indians of the South American Chaco even included the nose and ears. Scalps were commonly dried over smoke and mounted on wooden hoops.

It should be noted that not all heads preserved in tribal societies are trophies. Heads of ancestors are sometimes decorated in exactly the same way as those of enemies. Nor are all trophy heads kept intact or treated with respect. The skull may be broken and the pieces

35. Trophy heads are either discarded after being used for ceremonial purposes or kept as tallies of the enemy dead or for their continuing beneficial effect. Of the three skulls seen on the outer wall of this Ifugao house of Luzon, the lower jaw of one has been removed to serve as a handle for a gong.

36 and 37. (*Right and above right*) Message sticks decorated with incised pictographic symbols were used by Australian tribes as invitations to intertribal ceremonies. Their bearers enjoyed immunity in enemy territory and could combine their invitations with a proposal for peace. The sketch above shows the markings on the stick. *Prague, Náprstek-Muzeum*

38. (*Above, far right*) Tribal symbols for war and peace were adopted by Western civilizations in their dealings with tribes. Medals issued by the United States throughout most of the nineteenth century commemorating the inauguration of presidents show both peace pipe and hatchet. These and other medals were given to Indians at treaties or when visiting Washington. *3in (7·5cm). Vienna, Kunsthistorisches Museum*

distributed, as among some tribes of the Philippine Islands, or it may be used for a prosaic purpose – the inhabitants of the Melanesian island of Tanga use it as a lime container. Trophies other than heads were taken: the Abipon of the Chaco and the Maori made whistles or flutes from bones. The Chibcha of Colombia are reported to have preserved entire bodies, the skin stuffed with ashes and the face modelled with wax.

Peace-making

Even during a state of war, certain places or persons may be excluded from armed conflict, in accordance with the rules of conduct guiding the behaviour of the warriors. Among the Jalé of the New Guinea highlands the men's house of a village is taboo to the enemy, who otherwise burn and plunder, and the warriors will retreat to it for safety. On the Solomon Islands people in boats or climbing trees would not be attacked, on the New Hebrides certain paths were declared neutral ground. Besides these local restrictions on warfare, there were intertribal places of peace and refuge, such as the quarry from which many tribes of the North American Midwest obtained stone for the manufacture of their pipe-heads.

Persons considered sacrosanct could be those of high status, or messengers on peaceful errands. Chiefs were spared because their deaths would have to be compensated in kind, or because their greater

39. The rib of a palm leaf adorned with various feathers was offered by the Pacaas Novos of western Brazil as a first token of peace. Some of the tribes along the Guaporé River were only pacified during the 1950s. *60in (150cm). Vienna, Museum für Völkerkunde*

spiritual power meant that their ghosts were potentially more harmful. Messengers usually carried symbols of office which guaranteed them a safe-conduct.

Such messengers were ideally suited to initiate the peace which was the ultimate goal of wars arising from specific conflicts. They could be selected from a neutral community or from the ranks of those who by birth or marriage were affiliated to both warring tribes. The symbols conveyed by them to the opponents were different ways of spelling peace: the green branches carried by emissaries in many parts of Oceania referred to life; a bundle of bows and arrows tied together was used by the Lengua of the Chaco to signify the cessation of warfare; strings of shell money or other gifts of value promised compensation.

40. A pair of reeds decorated with eagle feathers and the skin of a duck's head were used by many North American tribes in ceremonies relating to peace, friendship and war. Fitted with a pipe bowl these became 'peace pipes'. In some dances they were employed to seal friendship by establishing ceremonial kinship, in others to invoke the powers of the war eagle to doom the enemy. These pipes also served as passports for ambassadors. *Oil painting by Paul Kane. Toronto, Royal Ontario Museum*

An evening of the scores was a pre-requisite for the restoration of peace. Compensation had to be paid for the loss of life or of property suffered by either or both groups. While the losers could be (and usually were) forced to make such payments, the victors might also offer gifts to their former enemies, beginning a series of exchanges of presents that would strengthen the new bonds by establishing mutual obligations.

Symbolic rituals likewise followed the conclusion of peace. Wampum belts woven in designs of white and purple shell beads commemorating the agreement were exchanged by tribes in the northeast of North America and kept in the tribal archives. Hatchets (among the American Indians) or swords (among the Tuareg) were

metaphorically or actually buried, and the peace pipe smoked. The tribes would feast, dance and drink together to cement their newly restored unity.

Victory feasts were also celebrated to honour the heroes, to mourn the dead, and to deride the enemy. If prisoners had been taken, they might be tortured after the return from the battlefield, enslaved, or adopted in place of relatives killed on the warpath. Trophies were proudly displayed and war deeds recounted.

Mock battles were sometimes staged as part of the peace-making ritual. Such fights served as a harmless way to release any remaining aggression before the resumption of peaceful relations, or to even the score, at least symbolically. They were also a way of allowing the real losers to gain compensation for actual losses during the war proper, since they were allowed to win the mock fight. The peace-making

41. Afraid of the superior military technology of Europeans, Tahitians humbly sue for peace by offering a green palm leaf. *From J. Hawksworth,* **An Account of the Voyages undertaken . . . for making Discoveries in the Southern Hemisphere,** *1773.*

42. Since wampum is a commonly used shell money in northeastern North America, the wampum peace ritual probably developed out of wergeld payments for injuries. The 'peace path' belt of the Huron Indians commemorates the initial proposal for peace between the Tobacco Nation and the Huron Confederacy whose three major council fires are indicated on the belt. *Length (excluding fringe): 22½in (57·5cm). Oxford, Pitt Rivers Museum*

fight of the Murngin was something of an ordeal: the aggrieved party threw spears at its opponents, who were not supposed to fight back. After their anger had waned, both groups danced together. Instead of settling the dispute, such mock fights would occasionally flare up into a resumption of full hostilities.

On the Solomon Islands both sides alternately staged mock attacks but stopped short of causing damage, while on the New Hebrides the actual losers, on winning the sham fight, would receive a payment of pigs for their victory. In areas as far apart as Melanesia and North America, mock battles have become a standard form of expressing friendship and are customarily used in welcoming visitors.

Readjusting to peaceful life may involve almost as many restrictions as going to war, because in both cases men are in transition between a normal and an exceptional phase. The spiritual powers needed and involuntarily encountered when facing the enemy may be dangerous in daily life. Many ceremonies held after the return from war are consequently aimed at the protection of the warrior against the vengeful spirit of his slain opponent. The successful warrior may take a new name, not only to remember his glorious feats but primarily to disguise his identity. Purification rites, such as the 'Enemy Way' of the Navajo of southwestern North America, were regarded as desirable for men who had come into contact with potentially dangerous aliens. The Enemy Way, for example, was a curative ceremonial held not only for the protection of returning warriors but also to treat Indians who had frequented white prostitutes.

Although most tribesmen can and do give detailed accounts of their warlike exploits, they frequently desire to create permanent, visible records of those achievements. These mementoes bear witness to the warrior's quest for prestige and status. They are proof of his increased social standing after the fighting is over, and ensure that his bravery benefits him in times of peace.

Trophies may serve as mementoes and although their ultimate purpose is usually different, they convey the message that their owner has scalped or beheaded his enemy. They are sometimes proudly displayed in or around the dwelling, or they may be carried around or attached to frequently used objects. Scalps with locks of hair still attached decorated many Plains Indian shirts; lower jaw-bones were in great demand amongst most tribes of the Philippine Islands as handles for their Chinese bronze gongs.

The most direct way of recording a warrior's deeds is to apply permanent marks to his body. Only those Kalinga warriors who participated in headhunting expeditions were privileged to tattoo themselves; they could extend this right to their female relatives. The

Dyak added a tattoo design to their body for every head taken. Among the Mohave of southwestern North America a scalp-holder might wear a distinctive tattoo pattern.

A highly developed system indicating specific war honours was employed by several North American tribes. A Dakota warrior who had killed an enemy wore an eagle feather with a red spot; if the feather was notched and its edge painted red he had cut the enemy's throat. Differently cut feathers indicated whether he had 'counted coup' or had been wounded in the battle.

Similarly, personal items could be adorned with symbolic decorations. Ermine skins attached to the shirt would identify a Crow Indian who had returned from the warpath with an enemy's gun. Registers of the number of battles a warrior had fought, horses he had stolen or enemies he had killed were painted on shirts or engraved on clubs and bows. Pictographic representations of particular feats adorned the shirts, bison-skin robes or tipi covers of victorious Plains Indian warriors. These paintings are among the most outstanding achievements of North American Indian art.

43. The scalp dance of the Hidatsa Indians was performed after a war party had returned with scalps. The women, partly dressed in men's costumes and carrying weapons, danced with the trophies attached to long poles, while the men sang songs mocking the enemy. On this occasion the scalper took a new name, perhaps to avoid retaliation by the ghost of the scalped. *After a drawing by Karl Bodmer*

44. One of several methods employed by the North American Plains Indians for publicly displaying war honours was by painting bison-skin robes or leather shirts with designs which were either stylized representations of battles or pictographic registers of individual acts of courage. *Berlin, Museum für Völkerkunde*

The Tribal Arsenal

Weapons may be used for either hunting or war, and it is often impossible to distinguish between the two functions. The problem is illustrated by the story of an attempt to disarm the natives of British North Borneo after the establishment of the colonial administration. To put a stop to war and headhunting expeditions but at the same time not to prevent the native Dyak from hunting, a list of all their weapons was to be compiled which would enable the administration to distinguish between the two types and enforce a ban on instruments of war. The project had to be abandoned after it became apparent that the Dyak recognized an almost endless variety of spears, only a few of which were obviously specialized for either hunting or war. The same problem applies to such weapons and tools as knives or axes. Native Australian women frequently used their digging sticks as improvised weapons in camp fights, and many of the so-called swords of tribal people may just as well be regarded as bush knives.

Weapons of war may or may not be employed in actual fighting. By association with the acquisition of power and status they tend to become symbols of rank and dignity. Such ceremonial weapons will frequently be of better workmanship, design, and overall artistic quality. Highly ornamented weapons intended for display rather than fighting may in turn become utterly useless as arms and degenerate into badges of office. Sceptres (derived from clubs) or heraldic blazons (derived from shields) are among the better-known examples of this process.

Each category of the tribal arsenal is discussed in the following pages. Just as the function of the various types of weapons significantly affects their form, so does the form influence the possibilities of artistic design.

Knuckle-dusters and fighting bracelets

The smallest hand-weapons used for striking are rings or bracelets armed with spikes or blades. Circular iron knives worn on the wrist or finger are known from East Africa and western Asia, while rings with one, two or several spikes have a somewhat wider distribution among African, Asian, and Indonesian tribes. In Oceania, some similar weapons are made from shark's teeth. According to an early description, 'the bowels or lower parts of the body were attacked with it, not as a dagger is used, but drawn across like a saw.'

Knuckle-dusters and fighting bracelets are rarely used in regular warfare. They do not extend the reach of the fighter and are more frequently employed in ceremonial boxing matches or to inflict punishment. Their main advantage is that they can be easily concealed. In the Society Islands, shark-tooth weapons were often used by women mourning: 'With this on the death of a relative or a friend,

45 and 46. A janus-faced head inlaid with shell separates the handle and the blade of the Guyana Indian club. The deeply incised decoration filled with white paint covering both sides is characteristic of the clubs of the area, while the weapon itself is an unusual type. Feathers were possibly tied to the little extension at the upper end. *19in (48·5cm). Stuttgart, Linden-Museum*

48. *(Above)* The weapons carried by the Ashanti warrior on this little gold weight from Ghana went out of actual use around 1700. But both swords and basketry or leather shields have survived as ceremonial regalia or badges of office up to the present day. *3½in (9cm). Berlin, Museum für Völkerkunde*

49. *(Above centre)* What looks like a bird's head is identified by the New Caledonian makers of this type of hard-wood club as a turtle's head. Being difficult to kill, turtles are sometimes employed as protective designs. However the pointed head suggests aggression rather than defense. *26½in (67·5cm). London, British Museum*

50. *(Above right)* Jivaro shields are usually painted with black figures of spirits, giant snakes and dangerous animals to inspire the enemy with fear and to increase the valour and strength of its bearer. This old shield exhibits stylized red and black painted designs. *35in (89cm). Vienna, Museum für Völkerkunde*

47. *(Left)* Of the two varieties of the Nago *dao*, one is a machete-like hewing knife while the other resembles a hatchet. The long socket of the hatchet and its open-sided scabbard are decorated with engraved geometric designs. The hewing knife is adorned with long tufts of dyed goat hair. *34¾in (88cm); 24¾in (63cm). Vienna, Museum für Völkerkunde*

they cut themselves unmercifully, striking the head, temples, cheek, and breast, till the blood flowed profusely from the wounds.'

Clubs

Of all the weapons used in tribal societies, clubs show the widest variation in form and offer the greatest variety of artistic design. A simple wooden stick was probably one of the first improvised weapons employed by man, but club design has advanced a long way from this prototype.

The shape of a club is primarily determined by its function as a striking weapon. Few ways can be found to improve the effectiveness of the basic weapon: the centre of gravity can be displaced towards the distal end, the striking edge can be made wedge-shaped, or the striking-power can be concentrated at one point. All clubs therefore expand to some extent towards the distal end. With this feature in common there are two types: one more or less flat, the other with a prominent head. Both may or may not be armed with additional blades or points.

The flat kind, variously described as a sword, sabre, paddle, oar or lath club, lends itself primarily to two-dimensional decoration, with incised, engraved or painted designs, more often ornamental than representational, covering parts of the surface. Three-dimensional carvings may sometimes occur (as on some short New Zealand clubs); they protrude from the edges and are necessarily rather flat. Most of these clubs are carved from wood, but some types of Maori *patu* and some northwestern American clubs are fashioned from bone or stone.

Globular or egg-shaped stone heads attached to wooden handles occur in Melanesia and North America. Maces with metal heads are more common outside the tribal world, but examples from the Sudan

51. *(Top)* A bracelet armed with a spike from Tanzania. 5¼in *(13cm). London, British Museum*

52 and 53. *(Above left and right)* The incised and slightly profiled fighting bracelet of the Afar of East Africa *(above left)* does little more than improve the effect of a blow with the fist. The spiked Shilluk bracelet, on the right, is apt to cause much more serious wounds. *Left: 3in (7·5cm); right: 3in (8cm). Hamburg, Hamburgisches Museum für Völkerkunde*

54. *(Right)* The natural fork of an antler is used to concentrate the striking power of this Ingalik club of northwestern North America into a point. *21in (53·5cm). Berlin, Museum für Völkerkunde*

55. *(Far right)* Heavy hardwood clubs of various shapes were used on the Tonga Islands. Those which had killed many enemies were engraved with figures, given proper names and regarded as supernatural beings. *Length: 43in (110cm). Exeter, Royal Albert Memorial Museum*

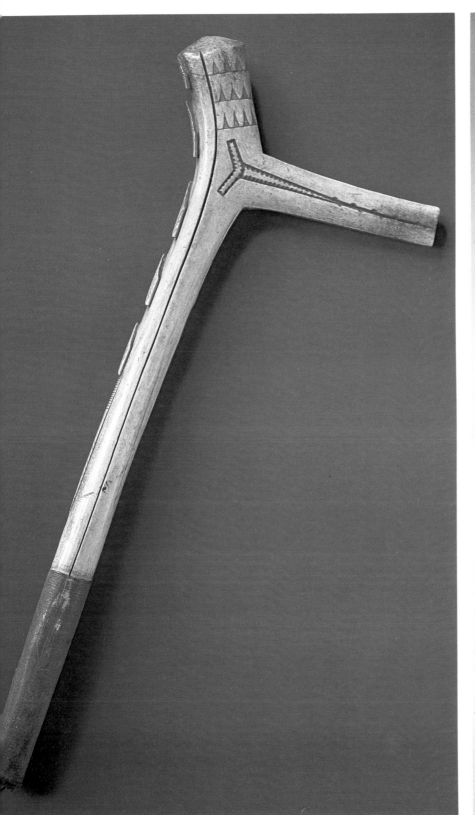

56. Wooden or plaited knuckle dusters set with shark's teeth were used in parts of Polynesia and Micronesia to tear open the antagonist's belly in man-to-man-fighting. Similar implements were used to cut up the bodies of slain enemies. This Hawaiian example was collected by James Cook in 1778. 6½in (16cm). *Vienna, Museum für Völkerkunde*

are known. The majority of the second type, however, are made of one piece of wood and many show exceptionally beautiful carving on the head, which indeed is frequently made to resemble an actual head.

The lower portion of most clubs is less distinguished. There is usually some kind of handle, and sometimes the proximal end is pointed so that it can be used for stabbing.

Although some clubs are used for hunting, most are specialized implements of warfare. For ceremonial use in war dances and the like, many of them are too heavy and unwieldy. Dance clubs are consequently often fashioned of lighter wood, or are smaller. On the other hand, where decoration takes precedence over function the weapon will become less efficient and less practical. The elaborately carved sword clubs of the Trobriand Islanders, for example, were carried by the men primarily for decoration and were very rarely employed for fighting.

Several types of club could be used for hurling at the enemy, including the specialized type with enlarged head.

57. Incised spirals and lozenges, partly filled with white paint, decorate a rare type of sword-shaped club from Brazil. Of two similar examples collected in the early 1800s, one is ascribed to the Paumary, another to the Mayoruna whose territory lies 700 kilometres farther west. 24in (61cm). *Munich, Staatliches Museum für Völkerkunde*

Battle-axes

The majority of axes were used as tools; battle-axes were rather unsophisticated weapons somewhere between tools and clubs. The blades in which the striking-power was concentrated were usually less well integrated with the shaft than in most comparable clubs. The major attraction of the battle-axe was the number of uses it had.

Few stone axes were employed as weapons because of their great weight. An important exception would seem to be the anchor-shaped

58. *(Far right)* Fiji, the easternmost Melanesian group of islands, has produced a great number of different types of clubs, among them both the heaviest and shortest of all Melanesian clubs. Some are inlaid with whale ivory, and all of them are well carved. The three-dimensional human head on this fine piece is unusual. 42in (106·5cm). *London, British Museum*

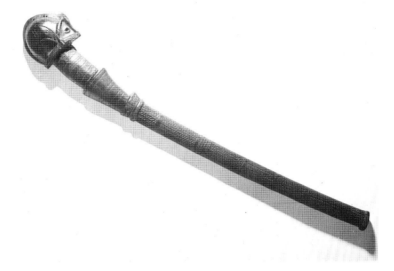

axe of the tribes of eastern Brazil, but even in this case, it was mainly carried by chiefs as a symbol of their authority. In Africa, where iron axes were widely distributed as wood-chopping implements and regularly carried around by the men, they came to acquire a secondary function as improvised weapons. Again, no specialized battle-axe type developed.

Axes did, however, acquire symbolic status through ceremonial use. A widespread West African belief connects the axe with the Thunder God (just as some North American clubs were regarded as instruments of the Thunderer, striking man like lightning). Here and elsewhere axes were employed in divination. They might signify the male principle, and consequently became symbols of dignity.

A specialized battle-axe was to be found among the tribes of northern Luzon, used for cutting off enemies' heads. The axe-head had a pointed extension on the back which was rammed into the ground when the edge was to be used as a knife.

Iron hatchets were introduced to North America in early colonial times as a tool, but almost immediately came to be used by the Indians as weapons as well. They partly superseded traditional wooden clubs, and were also thrown (a technique not otherwise known except in the case of some medieval European battle-axes). Combined with smoking pipes they became known as 'pipe tomahawks'. 'This', says an eighteenth-century British soldier among the Cherokee, 'is one of their most useful pieces of field-furniture, serving all the offices of hatchet, pipe, and sword; neither are they less expert at throwing it than using it near, but will kill at a considerable distance.' The hatchet soon became a metaphor for war. 'Burying the tomahawk and raising

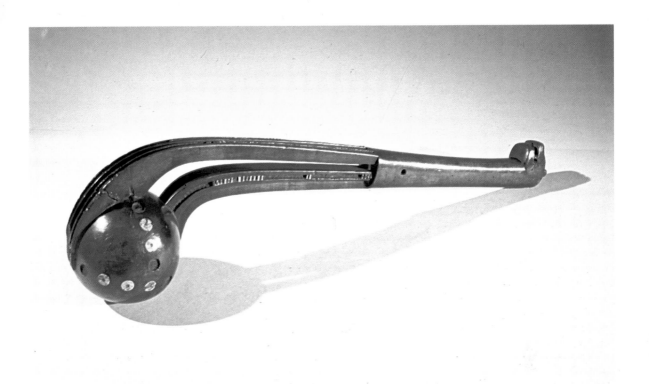

60 and 61. A ball-headed wooden club from the Eastern Woodlands of North America. It is inlaid with shell, and its lower end is in the shape of an animal's head. *London, Horniman Museum*

59. Besides several varieties of short wooden, whalebone or greenstone clubs of the *patu* and *mere* types, several kinds of long clubs were used by the Maori of New Zealand. Tamati Waka Nene, an important early nineteenth-century warrior chieftain of the Ngati-Hoa, carries a *tewhatewha* with a janus mask carved on the shaft. The attached bunch of hawk feathers served to divert the attention of the opponent. The blade is not used for striking but adds weight to the striking end. *Oil painting by Gottfried Lindauer. Auckland City Art Gallery*

63. An ingenious invention dating to about 1700 which greatly increased the popularity of iron hatchets among North American tribes was the pipe tomahawk. By attaching a pipe bowl opposite to the blade and piercing the handle lengthwise, the hatchet could be used for smoking, warfare and woodworking. Chiefs were usually presented with particularly ornate tomahawks. *12in (31cm). London, British Museum*

64. *(Above right)* The single-edged Mandingo sword from Mali has both its hilt and scabbard encased in leather decorated with ornamental incisions. *Sword: 28in (71cm). Berlin, Museum für Völkerkunde*

62. *(Left)* This anchor-shaped axe is one of the oldest eastern Brazilian weapons surviving to this day. Its original beauty can only be guessed since the rich feather fringe is almost completely gone. The ostrich eggshells with which the upper part is decorated must have been traded from the south, making it an extraordinary weapon even at the time of its manufacture. *25½in (65cm). Vienna, Museum für Völkerkunde*

a heap of stones thereon', 'fastening their hatchets to the sun', or 'hanging up the hatchet' were actions and expressions symbolic of making peace.

Swords

In a very broad sense the term 'sword' covers all metallic edged weapons larger than a knife. It may be used either for striking or for thrusting or both. In this it combines the advantages of a club and a dagger. Various types can be distinguished according to the shape of the blade (straight or curved) and whether it is single- or double-edged, pointed or blunt. Even if most tribal swords were all-purpose tools rather than specialized weapons, they could be 'altogether a most handy and formidable implement in the paw of a lusty naked savage', to use the words of a late-nineteenth-century British botanist who was summing up his experiences among the Dyak.

A great number of swords used in tribal societies were really hewing knives, the function of which was to cut by striking. The width or thickness of the blade was therefore generally increased towards the distal end, in order to fix the centre of gravity in the lower half of the blade. Like axes, these swords commonly also served as tools. At the same time, there is no doubt that their primary purpose

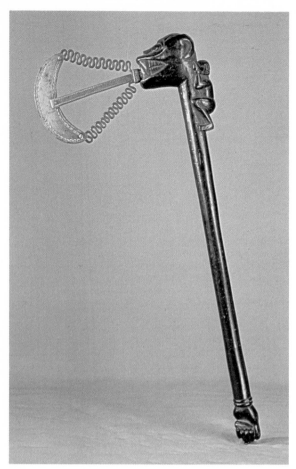

65. In West Africa, axes frequently take the function of divine symbols. The half-moon shaped axe blade, for example, represents lightning emanating from the mouth of the Dahomean thundergod Xevioso. As kings were regarded as incarnations of deities, axes came to symbolize royal power and the king himself. *Paris, Musée de l'Homme*

66. An eighteenth-century Nootkan club of the type thought to have been used in the ceremonial killing of slaves. *11¼in (28·5cm). London, British Museum*

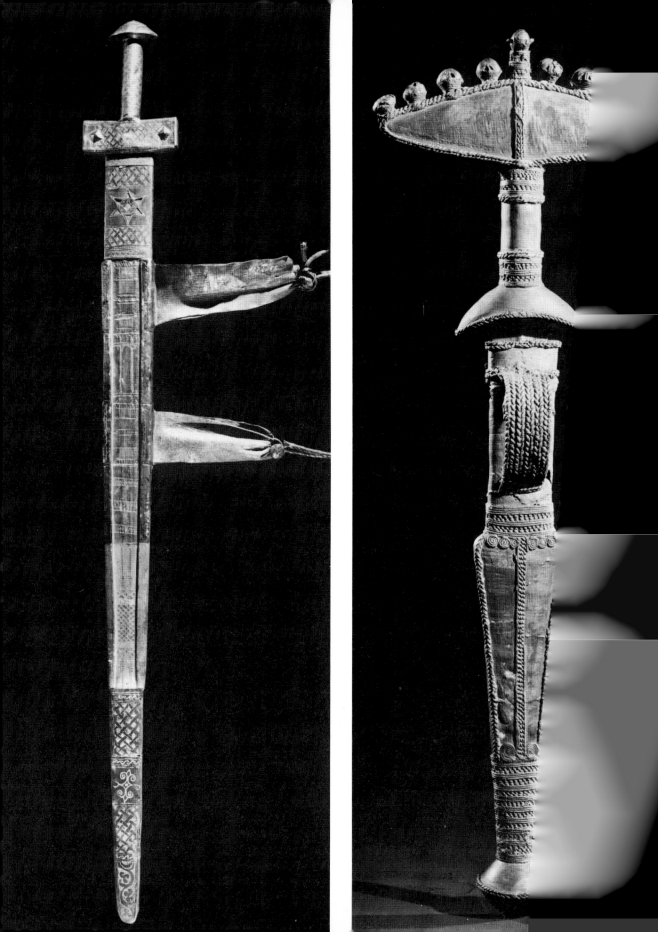

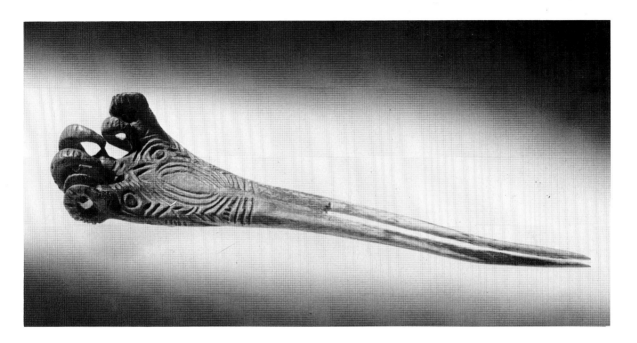

69. In many parts of New Guinea, carved cassowary bone daggers were restricted to successful headhunters, the cassowary being regarded as a symbol of swiftness and strength. These daggers were used for fighting and stabbing pigs, because pigs were sometimes substituted for human victims in a ritual equivalent to cannibalism. The double point makes this specimen from the Abelam of the Maprik area unusual. *10in (25·5cm). Missouri, St Louis Art Museum*

67. *(Far left)* The long, straight *tacouba* is the major weapon of the Tuareg of the Sahara. Having become part of the male dress, it has survived all other Tuareg arms. The design on the pommel refers to the shield – a magical device believed to be helpful in splitting the enemy's means of defence. *38⅜in (98cm). Vienna, Museum für Völkerkunde*

68. The iron blade of this Batta dagger from northern Cameroon is attached to a highly ornamental brass hilt. The scabbard is also made of brass. *Dresden, Museum für Völkerkunde*

was that of a weapon of war. The headhunting weapons of the Naga of Indonesia and the Philippines, as well as some African swords, belong to this group.

Straight, pointed swords with two edges were much less common. Wherever they did occur (for instance in Indonesia or northern Africa) they appeared to result from the influence of neighbouring civilizations, even though it is quite possible that they could have developed from oversized daggers. Tuareg swords, for example, were originally armed with German blades. As with most metal weapons, swords were generally the work of specialists. Good blades were often widely traded, to be hafted by their ultimate users. They were sometimes decorated in accordance with the customer's taste.

Knives and daggers

Smaller single-edged knives served predominantly as cutting tools. Besides being used occasionally as weapons, they were also useful in warfare for scalping or similar purposes. For actual fighting, double-edged and/or pointed daggers had a much wider distribution because they were specialized stabbing weapons. In their simplest form they took the shape of plain wooden sticks sharpened at one end. Such daggers are reported from Australia and Hawaii. Similarly, splinters from long bones made simple but effective weapons. Stone blades with wooden handles attached usually served other functions as well. Scope for decoration was largely restricted to the hilt.

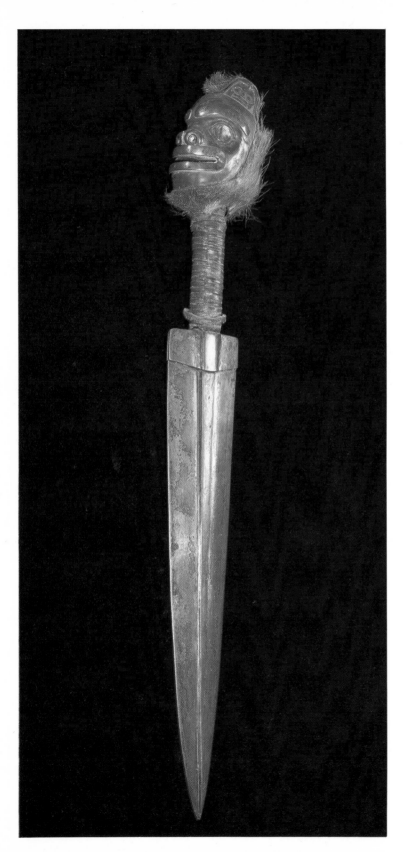

70. *(Left)* Double-bladed daggers of iron or native copper used by the Tlingit and other tribes of the Northwest coast of North America were largely replaced by single-bladed daggers whose blades were of European origin. This late nineteenth-century piece combines a copper blade with an exquisitely carved and inlaid haft representing a mythical bear. *Cambridge (Mass.), Peabody Museum*

71. The carved wooden pommel of this Dyak *Malat* or *mandau* ('headhunter') from central Kalimantan combines a stylized human figure with dog- and leech-motifs which symbolically refer to headhunting. The dog is used as a substitute for the mythic tiger to which open reference is avoided. Human and goat hair is attached to the hilt. The blade is curved and will cut only when a certain technique is employed. A smaller knife is always kept in a small sheath attached to the main one. *Length of sword: 29in (74cm). Leipzig, Museum für Völkerkunde*

Copper daggers were found in northwestern North America, while iron daggers of tribal origin occurred primarily in Africa. Weapons like the *kris* and the *katar*, found in Asia and Indonesia, were the product of higher civilizations, but they occasionally found their way into tribal use, just as European knives were adopted by American Indian groups. Special types included the looped daggers of West Africa and India, the hilt of which completely encircled the hand, and the arm daggers which were carried in sheaths attached to the arm rather than the belt. Metal blades were often curved rather than straight and were sometimes decorated.

Lances

Lances are hand-held poles designed for fighting at close quarters. They resemble javelins in that they are piercing weapons, but they lack stabilizers which would balance them in flight. Their shafts are rigid and tend to be thicker at the lower end.

Some lances, such as those of Wuvulu and the Gilbert Islands, have additional blades or points for striking the enemy and show a marked similarity to the pole arms of Eurasia. Other types of lance are less specialized and may be used for throwing as well.

Spears and spearthrowers

The borderline between spears and arrows is as indistinct as that between lances and spears. Although spears are usually longer and lack feathering, there are actually arrows to which these criteria apply. The only reliable distinction is that arrows are propelled by bows whereas spears are thrown by hand or occasionally with the help of spearthrowers. Spears may be stabilized in flight by feathering, but more generally by displacing the centre of gravity towards the front end. This is done by adding heavy projectile points, wooden shafts (on reed spears), or simply increasing the thickness. Most of the decorative features will therefore be concentrated at or near the point. There are sometimes carvings on the shaft near the haft; the point itself may be decorated or fitted with barbs, which besides their practical use frequently show excellent workmanship and artistry.

The hope that supernatural means may affect the performance of the weapon occasionally influences its construction and design. The Jivaro, for example, selected chonta-palm wood for their spears because they believed that this palm was the seat of a spirit. The lower end of some New Guinea spears is embellished with a socket of human bone to endow the weapon with special powers.

Spearthrowers were employed to gain better leverage for throwing. A few were flexible, like the New Caledonian throwing cord which was wrapped around the shaft, but most were wooden levers with

73. Two early nineteenth-century Mundurucú spears exhibit geometric painted designs on the concave side of their broad bamboo points. The shafts are elaborately decorated with fur, feathers and cotton thread. The third spear with the carved and inlaid face below the point was captured from the neighbouring Parintintin. *60½in (154cm); 54½in (138cm); 60½in (154cm). Vienna, Museum für Völkerkunde*

either a socket or peg or a combination of both holding the projectile. In recent times the spearthrower was of limited distribution, found only in Australia, parts of Oceania and some areas on the American continent. Some of the Australian types were painted or incised, while in New Guinea and the Americas carved decoration predominated.

Bows and arrows

Since the bow undoubtedly represents the highest development of arms technology in the tribal world, it seems strange that it is not always employed as a weapon of war. In Polynesia bows and arrows were restricted to hunting; in parts of Melanesia the spear replaced the bow, and even the civilizations of Mexico and Peru preferred the spearthrower. Since there are no technical reasons for this, it is likely that the bow was less suited to the particular war tactics of these regions.

Bows vary immensely in shape and construction. The effectiveness of the simple wooden bow in warfare is proved by the medieval English longbow, and many tribes use this type. Length is essential for power. Mounted warriors, on the other hand, need shorter bows for the weapon to be practical when riding. The bow staff can be reinforced by layers of sinews glued to its back, but while effective, most glued bows weaken under humid conditions. Maximum power is attained by the laminated bow of Asia which consists of various layers of wood, sinew or cloth, and horn. Decoration of the bow staffs is usually restricted to painted designs on the back, and occasionally on the underside. Most carved and wrapped decorations tend to weaken the bow.

Hunting arrows and war arrows differ to some extent. Hunting arrows should be easily removable from animals so that they may be used again; war arrows, on the other hand, tend to have barbs to make removal by the enemy difficult. Sometimes part of the shaft is detachable, so that the head will remain in the wound when attempts are made to pull out the whole arrow. Because human ribs are horizontal and the bow is usually held vertically, the head of a war arrow is often perpendicular to the plane of the notch, so that the point may pass more easily between the ribs. For the same reason, the plane of the head matches that of the notch in hunting arrows. Arrows are sometimes painted, more frequently on or around their feathering than anywhere else. Some of the designs are apparently property marks.

The art of archery has noticeable stylistic variations: for example, five distinct methods of releasing the arrow are practised. In the primary method the butt end of the arrow is grasped by the thumb and index finger; in the secondary and tertiary types both the arrow

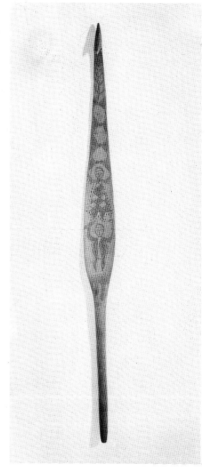

73. The lower figure depicted on this engraved southeastern Australian spearthrower carries a boomerang in his right and a *tjurunga* in his left hand. The latter is a sacred object of wood or stone, used in rites to increase and perpetuate the fertility of nature. *London, British Museum*

74. Despite Lukas Cranach's pictorial report from Paradise, it is unlikely that the First Archer (let alone the First Man who certainly knew nothing about the bow) practised the Mediterranean method of arrow release. This release which is known to only some tribal populations is said to enable the most powerful pull and was probably the result of long experience. *Oil on canvas. London, Hampton Court Palace*

75. Three Melanesian spears illustrate some
of the wide variety of forms and designs
encountered in this simple weapon. The
elegantly shaped asymmetric barbs are
carved on the wooden point of a spear from
the Torres Straits Islands, while they are
made of bone and attached with cord on
the Solomon Island spear. The spear from
the Admiralty Islands features the typical
obsidian point but is unusual for having
two of them. *63¾in (161cm); 105in
(265cm); 55in (140cm). Vienna, Museum
für Völkerkunde*

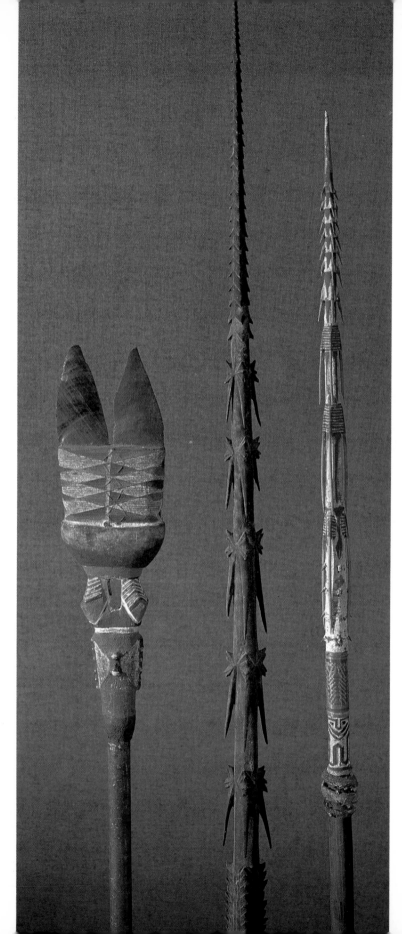

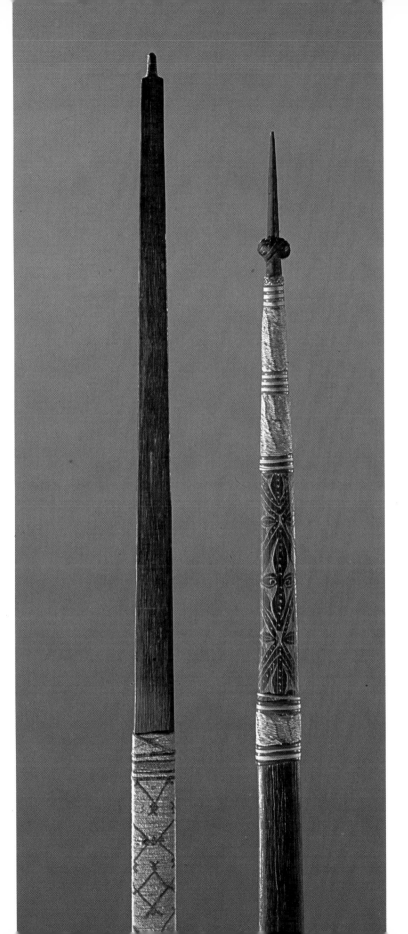

76. Besides being decoratively wrapped
with rattan and split palm leaves, the bow
from New Guinea *(left)* bears shallow
engravings of stylized faces on its belly. The
staff of the Conibo *(right)* from the Peruvian
headwaters of the Amazon is almost
completely wrapped with cotton thread
which has been painted in a design usually
seen on pottery and textiles of this area.
*72½in (184cm); 85in (216cm). Vienna,
Museum für Völkerkunde*

and the string are pulled in different ways. In the Mediterranean and Mongolian styles only the string is drawn: the latter method employs the thumb only (sometimes with the aid of special thumb-rings), while the Mediterranean method on the contrary does not use it at all.

Pellet bows, with which pellets rather than arrows are shot, are found in parts of Asia and South America but are never used for war. The crossbow has only occasionally been used in tribal warfare.

Slings and blowguns

As a hunting weapon and toy the blowgun is widely distributed, but concentrated particularly in Indonesia and northern South America. As an implement of war it is limited to Kalimantan and some other areas of the Indonesian archipelago, and is always the same type: a plain hardwood pole, bored and regularly fitted with an iron bayonet and sometimes a sight. The tiny darts, with only breath to propel them, would cause little harm to man were they not tipped with one of several poisons such as *ipoh* or *siren*. Even so, their use in war declined during the centuries of European contact.

While the blowgun is primarily suited to a forest environment, the sling is found to be more effective in open country. And while the use of the blowgun seems to have spread relatively recently, the sling is an ancient weapon, found on all continents except Australia. Literary and archaeological evidence confirms its existence several thousand years ago. It must have developed as a refinement of throwing stones by hand, a primitive method of warfare used until recently by some

77. This lithograph shows Dyaks attacking British expeditionary boats with poisoned blowgun darts which proved ineffective as they do not penetrate cloth at a distance of forty yards. Blowguns were no match for firearms, as the fact that they are noiseless was no advantage in open fighting. *Lithograph from R. Mundy,* **Narrative of Events,** *1848*

78. Jean Baptist Cabri, a runaway sailor naturalized on the Marquesas Islands, demonstrates the use of the sling (the sides are reversed due to the process of engraving). The sling passes behind the back and is manipulated by a quick pull of the right hand. When not in use, it was wrapped around the head. Tattooing of the breast was thought to ward off spears and bullets. *Engraving from G. H. um Langsdorff,* **Bemerkungen auf einer Reise van die Welt,** *1812, after a drawing by Orlovsky*

80. *(Above)* **To their neighbours the Dinka of Sudan are known as 'people with sticks' because they preferred to fight with wooden clubs even though the lance had become their major weapon. They also had special bow-like shields for parrying clubs. While most of their clubs were of the knobkerry type, this one is remarkable for its carved head.** *Vienna, Museum für Völkerkunde*

81. *(Above right)* **Three nineteenth-century boomerangs for use in war and hunting.** *From left to right: 22½in (57cm); 20in (51cm); 20in (50cm). London, British Museum*

79. **For the tribes of northeastern New Guinea, headhunting and cannibalism are a partial re-enactment of the struggle between male and female principles in the mythic past, deemed necessary for the maintenance of the cosmic order. Wooden shields from the lower Sepik area are occasionally decorated with a male figure in feather mosaic, which represents an aspect of this myth. These shields were not used in battle but were kept as sacred symbols in the men's house.** *Berlin, Museum für Völkerkunde*

tribal populations. The sling itself is a simple device consisting of a string with a pouch holding the sling-stone near its centre. Rotating the sling launches the missile. The scope for artistic design is limited. Most slings are made of leather or woven material; highly decorated pouches are mainly found outside the tribal world, such as the polychrome tapestry ones in the regions of the ancient Andean civilizations.

Throwing-sticks

Unworked pieces of wood and straight sticks pointed at both ends were used occasionally as missiles for hunting but scarcely ever in war. More sophisticated shapes were used as weapons.

Throwing-sticks are usually flat and either curved or angular. Their wedge-shaped edges inflict wounds regardless of which part hits the body. Those with definite handles are equally useful as hand-held weapons, and even throwing-sticks without handles are sometimes used for striking. Although throwing-sticks are widely distributed all over the world, the Australian boomerang is the best-known type. Boomerangs come in many different shapes, reflecting both regional variation and differences in function. Not all are used for fighting.

There are only three types which will return to the thrower, and none of them is a weapon. The beaked boomerang, on the other hand, is admirably suited to circumvent shields. Because it is regarded as a dangerous weapon its use is restricted to the leaders and older men, who use this 'peace-maker' to break up fights. The restriction means that it also becomes a symbol of rank. Boomerangs frequently feature in the mythology of Australian tribes and many of their uses are ceremonial. Quite a few are engraved or painted with abstract designs.

Throwing-knives

An extraordinary African weapon which has developed out of the throwing-stick is misleadingly called a throwing-knife, although it does not resemble a knife and is only occasionally thrown. From a straight or slightly curved central iron blade protrudes a variety of knife-, dagger-, sabre- or sickle-shaped extensions. A comparatively simple weapon of this kind, which is actually used in fighting by tribes of the central Sudan, seems to have been the prototype from which the grotesquely shaped and ferocious-looking types to the east and south are derived. Its many sharp points and cutting edges could cause serious injuries – if it were to be thrown at all. Most tribes in the area of its highest formal development, particularly in the Congo basin, were obviously reluctant to risk the loss of their precious weapon by throwing it away. Except when needed as a last resort it was rather carried as a threatening ornament or as a sign of dignity. In some cases it developed into a form of currency or was elevated to the rank of a fetish. In this trend from the practical to the ceremonial the throwing-knife follows the fate of European pole arms.

A throwing-knife of a different kind is the chakram, used by the Sikh of northern India. A disc with extremely sharp edges and a central hole, it is either rotated around one finger before being released or thrown from between thumb and index finger.

Firearms

Tribal people have always been quick to adopt for their own purposes technologically superior weapons introduced by outsiders, such as colonizers. Early colonial reports graphically describe the terror and amazement of tribesmen at their first encounter with firearms. But the initial timidity soon gave way to the desire to add these weapons to their stock. Tribes using firearms always depended on traders for their supply of arms and ammunition and were rarely provided with the latest models. While this put them at a disadvantage *vis-à-vis* their suppliers, the possession of firearms could make all the difference in intertribal warfare.

82. This Fang throwing-knife from Cameroon represents an intermediate stage in the development from the simpler Sudanic F-shape to the more ornamental forms. Fine lines engraved parallel to the edges stress the aesthetic character of the outline. *Dresden, Museum für Völkerkunde*

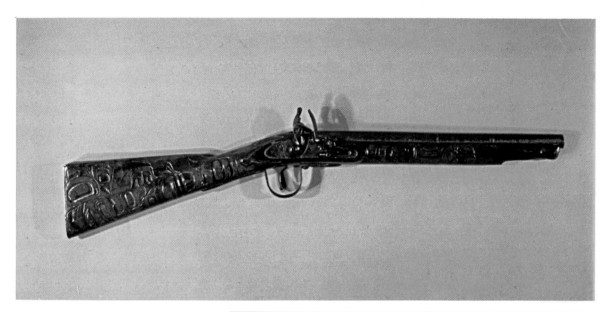

84. This Parker and Field flintlock musket has throroughly adapted to the taste of its nineteenth-century owner, a Haida Indian of British Columbia. The barrel has been shortened and most of the gun covered with the stylized representations of zoomorphic heraldic devices. *Vancouver, University of British Columbia Museum of Anthropology*

85. *(Right)* Pieces of leopard skin, an emblem of courage, are used both as an upper garment and in the making of the trousers of the native warrior represented by this solid statuette from Benin. His weapons include a European flintlock gun and a short, dagger-like sword, while at his feet lies a replica of a trophy head. *17¾in (45cm). Liverpool, Merseyside County Museum*

83. A Fang throwing-knife representing a bird's head. *12in (30·5cm). London, British Museum*

Although produced outside the tribal world, guns did become tribal weapons. They were sometimes held like a bow, with the extended left hand grasping the barrel, and consequently marksmanship was rather poor. From early nineteenth-century Polynesia we learn how the natives handle their firearms: 'in the most awkward manner, holding it above the head, or by the side, and in this singular position fire it off. I was once with a party of natives, when one of them fired at a bullock but a few yards distant, and missed it.' A valid explanation for this poor record is supplied by another nineteenth-century observer in Indonesia: 'This inexpertness arises from the want of practice necessarily incident to the scarcity of arms and ammunition, and to the practice of pursuing the chase by other means.'

By decorating the stocks of their guns in regional style, and by keeping them in cases fashioned like quivers, the new owners adapted their firearms to suit their needs. Metaphors such as the Tibetan expression 'fire arrow' for gun indicate how the weapons were integrated into the arsenal.

The effect of such weapons was far-reaching: traditional protective devices lost much of their former value, and since the recharging of muzzle-loading guns was time-consuming, new strategic patterns had to be developed. The net effect, however, depended on whether only one or both warring sides possessed guns. If only one tribe had them, it could use its superiority to push back or wipe out its enemies, and clear evidence exists that this happened in many part of North America. With both sides equally armed, both would suffer heavy losses. As the primary goal of each side was to minimize its own fatalities, the arming of both tended ultimately to lower the incidence of armed conflicts. The abandoning of headhunting and the establishment of intertribal peace pacts on the island of Luzon are largely the result of this 'balance of power' effect.

Catching devices

When the main purpose of a raid is to catch rather than kill the opponent, special catching devices may be used instead of weapons. Some headhunting tribes of New Guinea employ a horrible combination of a rattan loop and a spike for this purpose. No less effective as a man-catcher is the two-pronged fork set with barbs used by the Lanao of Mindanao. Lassoos were occasionally used in war or in slave-hunting in India, Sri Lanka and by the Arabs of the Sudan. According to tradition, the aboriginal inhabitants of Easter Island casts nets over their enemies.

Entanglement of the enemy's body and limbs was likewise the major purpose of the bola (or, more correctly, boleadora), used by the Indian tribes of southern South America. The bola-stones were

86. Already removed from the tribal tradition of anonymous art, the damascened barrel of this Afghan blunderbuss is signed by its maker. The main design of the stock is a stylized tree of life inlaid with bone, mother-of-pearl and brass wire. *23in (58cm). Vienna, Museum für Völkerkunde*

87. A late nineteenth-century illustration shows the use of man-catchers for headhunting by the tribes in the Gulf of Papua area of southwestern New Guinea. Most other details of dress, ornament and arms, however, are somewhat fictional. *Engraving from J. Chalmers and W. W. Gill, New Guinea, 1885*

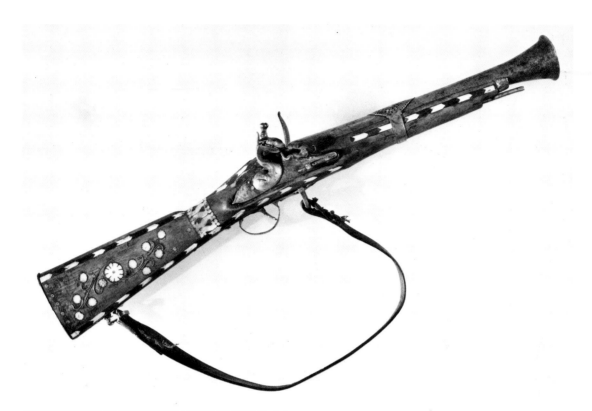

91. *(Right)* Quilted cotton armour protects both Kanuri warriors and their horses of the kingdom of Bornu in northeastern Nigeria in this late nineteenth-century illustration. *From P. Paulitschke,* **Die Sudanländer,** *1885*

fastened to each end of a rope, or to three ropes tied together; they could well cause bruises or concussion while tripping up the victim.

Armour

Normal clothing, worn to protect against extremes of temperature, is generally insufficient to guard a man against his enemy's weapons. The natives of the mountains of northeastern New Guinea overcome this by piling on up to twelve of the barkcloth capes which they usually wear. Although effective, this kind of rudimentary armour is cumbersome, and many tribes have tried to develop less inhibiting means of protection. Armour is particularly useful when the warrior needs both hands to make use of his weapons and cannot therefore carry a shield as well. It protects the body, or important parts of it, from all sides. Simple types consist of cylindrical bark or tough skin 'corsets' for the trunk, either worn as broad belts or hung from the shoulders. Examples from Melanesia, Africa and South America are sometimes decorated with geometric designs.

Tailored skin armour was used on both the American and Asian sides of the Bering Straits, and in Indonesia, where it was frequently reinforced with small metal plates. Woven rattan and coconut-fibre armour was also made here, as well as in Micronesia and New Guinea; its most highly developed form is found on the Gilbert Islands. The prominent neck protection which is a feature of Gilbertese armour is said to have been necessary to shield the warrior from stones thrown by the women from behind.

Wooden slat and rod armour is found in various parts of North America, and decorated examples survive from the Pacific Northwest which may ultimately derive from Chinese models. Metal is used rarely outside Old World civilizations, but imitations of metal armour were made in Siberia from bone and ivory. Islamic and European influence spread the use of coats of mail to the African Sudan and

88. *(Far left)* The central portion of the Ata armour from southern Mindanao strongly resembles the shape of the shields used in this area (compare plate 97). The armour itself is made of rattan, coated with resin, and embroidered with cowrie shells. The rattan helmet is crowned by a bunch of feathers. *Leiden, Rijksmuseum voor Volkenkunde*

89. Collected on James Cook's third voyage in 1778, this oldest surviving wooden slat armour from the northwest coast of North America shows vestiges of stylized painted decoration. The series of carved mask-like faces on the boards make it an outstanding piece in spite of the poorly preserved painted design. *Cambridge, Museum of Archaeology and Ethnology*

90. *(Left)* Reminiscences of the Middle Ages are aroused by warriors from Rei Buba in northern Cameroon clad in chain mail. When the Fulani established the empire of Adamawa during the early nineteenth century, they brought with them many artifacts of ultimately Mediterranean origin.

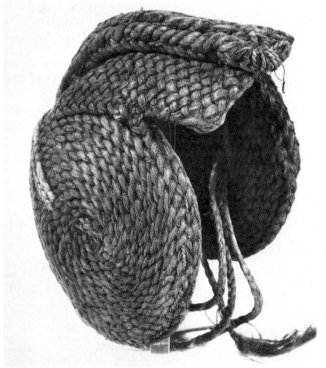

92. Tlingit wooden helmets consisted of two parts. The lower section or visor, with shallow notches along the upper rim serving as eye-holes, was topped by a mask-like helmet. The realistically carved human or animal faces, which were set with hair and opercula shell for teeth, added to the illusion of the giant size and ferocious nature of the warrior. *Berlin, Museum für Völkerkunde*

93. *(Left)* A light basketry war cap from Northern Cameroon shows a very simple type of helmet. A braided ridge gives added protection to the crown of the head. *8in (20cm). Braunschweig, Städtisches Museum für Völkerkunde*

94. *(Right)* Dreams and visions furnished the designs painted by the Plains Indians on their skin shields and the leather covers enveloping the shields when they were not in use. Despite the individualistic origin of the design, the same elements and compositions occur on many shields, probably due to their proven protective quality. A pair of birds and celestial bodies appear on this mid-nineteenth century Osage example. *17¾in (45cm). Offenbach, Deutsches Ledermuseum*

Indonesia. Quilted cotton armour is used by some tribes, again especially in the Sudan.

Some designs used in body painting were thought to give the warrior magical protection. In the same way the North American Plains Indian followers of the Ghost Dance, a late-nineteenth-century movement aiming to restore traditional ways, painted their cotton or skin shirts with protective designs similar to those applied to shields. It was believed that these shirts would even repel bullets.

Helmets not only protect the head but generally serve to make the warrior appear taller. Made of wood, skin, basketry or metal, they are frequently pointed or ridged at the top to soften or deflect blows. Helmets, being highly visible, invite adornment, and the feather helmets of Hawaii and carved wooden helmets of the Northwest

95. To protect themselves against the ugly wounds inflicted by their shark tooth weapons (particularly the long wooden spears and pole arms), warriors on the Gilbert Islands used to wear heavy armour made from coconut-fibre and helmets either of the same material or cut from globefish skin. Being restricted in their movement and easily tired by the weight of their protective garment, an assistant had to direct and hold the fighter from the back.

96. Only a small metal buckler protects the hand of the warrior on the Punjabi *márú*. Besides the shield's possible use as a parrying device, the metal-tipped antelope horn extensions and a central spike made it a dangerous weapon as well. *Hamburg, Hamburgischer Museum für Völkerkunde*

Coast Indians represent notable artistic achievements. That beauty sometimes took precedence over utility may be gleaned from a nineteenth-century description of the *fau* of the Society Islanders: 'This was also a cap, fitted closely to the head, surrounded by a cylindrical structure of cane-work, ornamented with the dark glossy feathers of aquatic birds. The hollow crown frequently towered two or three feet above the head, and, being curved at the top, appeared to nod or bend with every movement of the wearer. This was a head-dress in high esteem, and worn only by distinguished men, who were generally sought out by the warriors in the opposing army. To subdue or kill a man who wore a *fau*, was one of the greatest feats.'

Shields

The most widespread protective device employed in the tribal world was the shield. In its most elementary form it was a stick to ward off the opponent's blows; some Australian tribes used a broad spearthrower. True shields, however, always featured at least a rudimentary guard for protecting the hand and enabling it to grip firmly.

Types of shield developed along different lines modified by the particular mode of fighting and the degrees of protection and mobility required. Light hand-held shields were better suited for man-to-man combat with hand weapons, and for mounted warriors, whereas tall shields set on the ground offered more effective defence in pitched battles. Special devices were required for archers, who needed both hands to discharge their arrows. Depending on the technological skills and raw materials available, shields could be made of wood, bark, basketry, leather, or metal.

The surface of the shield is an obvious area for embellishment, and because the shield protects its bearer, many of the decorative designs also have protective properties. Consequently the shield is often regarded, as among the Jivaro, 'not only as a weapon of defence, but also as a religious instrument. . . . It is full of magic power, and when it is beaten with the lance in a particular way, this is in fact a sort of religious or magical ceremony which is believed to inspire the spirits – first of all the spirit of the slain enemy – with fear.' The Dakota Indians of North America believed that the supernatural powers called up by the painting on their shields exerted an influence on the enemy, causing him to shoot at the shield rather than at the exposed parts of the body. This reasoning certainly has a sound basis in that a conspicuous design will naturally attract the opponent's eye.

For this very reason only the most courageous of the Trobriand Islanders would paint their shields as an open challenge to the enemy. To split such a shield and kill its bearer was considered especially

97. The raised central section of this carved wooden Bagobo shield from the Gulf of Davao area of Mindanao is related to the construction of the handle on its back and may be compared to the circular shields used by the same tribe. No meaning can be assigned to the geometric decoration which sometimes is continued on the reverse side. *Leiden, Rijksmuseum voor Volkenkunde*

98. The outside of many Dyak shields is painted with one or more demons' heads, a design which possibly derives from Chinese dragon- or tiger-shields. Human hair surrounds the fearsome faces. The inside may be decorated with similar motifs or with a pair of smiling anthropomorphic figures. *Leiden, Rijksmuseum voor Volkenkunde*

99. *(Right)* The reverse side of the Dyak shield shown in plate 97.

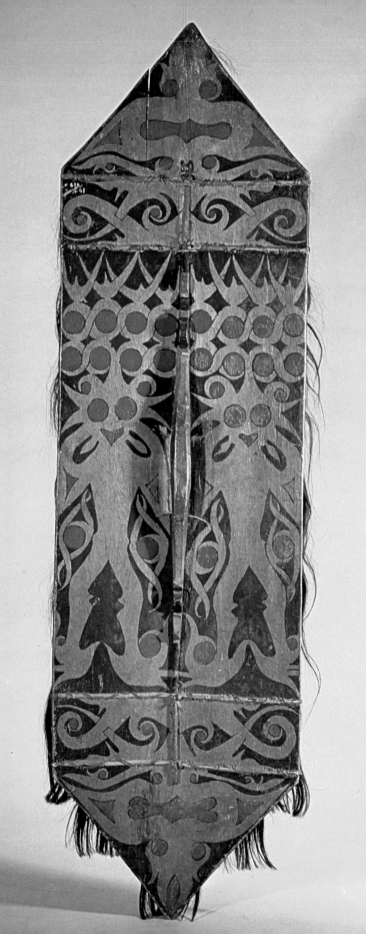

100. When threatened by raids of their slave-hunting Bagirmi neighbours, the Gaberi of southern Chad withdrew to fortified tree dwellings, taking some of their livestock with them. While they could pelt their besiegers with missiles from above, the safe harbour frequently turned out to be a trap when the tree was set on fire. *From Gustav Nachtigal,* **Sahara und Sudan,** *1881*

honourable. The introduction of firearms obviously relegated the shield to insignificance, partly because – as the Dakota noticed – it continued to attract the bullets like a target without actually holding them up.

In Melanesia, many of the elongated shields were decorated with more or less stylized human faces or bodies; in some cases even the outlines were clearly derived from the human figure. Indeed, such shields were frequently identified with ancient spirits, given the names of deceased relatives, and used in raids to avenge the death of the ancestor whose name they carried.

Among some tribes, on Kalimantan and New Britain, for example, shields were decorated both on the inside and outside, which indicates that these designs were addressed not only to the enemy but to the bearer as well. Other designs of a more abstract character identified the bearer as the member of a certain social or military unit or indicated the status he had achieved by his acts of bravery. In some regions such as East Africa definite heraldic systems were developed. Long before they fell into disuse as a means of defence, light dance shields and elaborately decorated ceremonial shields played a large part in tribal life. After the introduction of firearms, all that remained of the shield's functions were these heraldic and social aspects.

Fortifications and traps

The continual threat of sudden attack moved most tribes to take precautionary measures. Settlements might be located in easily defensible places offering strategic advantages, such as a hilltop or river-bend. Stockades or walls might surround the entire village or its most vital sections, and even the houses might be constructed with defensive needs in mind. Sometimes special structures, such as tree houses, were built, into which the villagers could withdraw when overrun by the enemy. The Jivaro erected square war towers up to 40 metres high, made of wooden poles, and somewhat smaller towers were built in the New Guinea highlands; they were permanently manned by warriors on the lookout for enemies trying surreptitiously to invade their territory. In some regions, particularly those where headhunting was practised, booby-traps were often set: trenches or pits were dug in the ground, lined with spikes and camouflaged so as to be invisible. The enemy might also unwittingly step on a mechanism which forcibly released a bent pole armed with sharpened sticks. The Naga and Dyak made use of bamboo calthrops which were likewise intended to hold up the enemy.

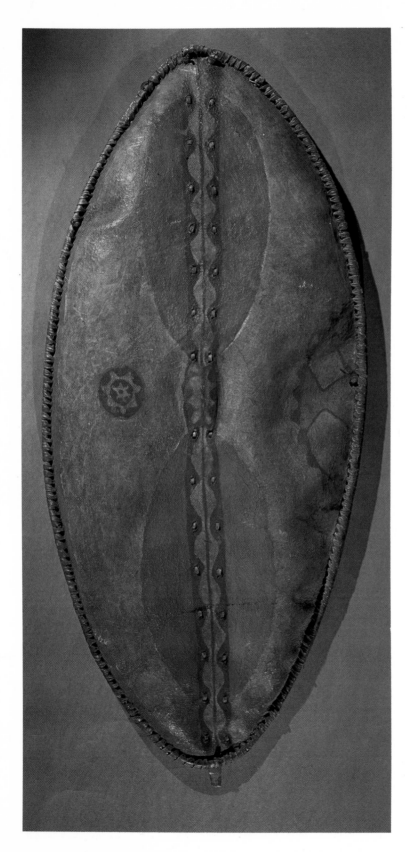

101. *(Left)* The leather shields used by the Masai of Kenya and Tanzania were painted according to a complex system of symbolic signs indicating status, age class, local group, fighting unit, kinship group and former military achievements. A vertical wooden bar, with a carved handle, and a wrapped wooden rim strengthen the shield. *Prague, Náprstek-Muzeum.*

102. *(Right)* Several examples survive of this type of oval basketry shield. It is decorated with figurative inlays of mother-of-pearl on a red and black coloured vegetable gumbase. Probably not made for combat, these shields may be the work of one mid-nineteenth century artist from Guadalcanal, one of the central Solomon Islands. *Vienna, Museum für Völkerkunde*

103. *(Far right)* A Solomon Islands ceremonial shield, probably from Isabel, where shields were rectangular in form. *Length: 30in (76cm). London, British Museum*

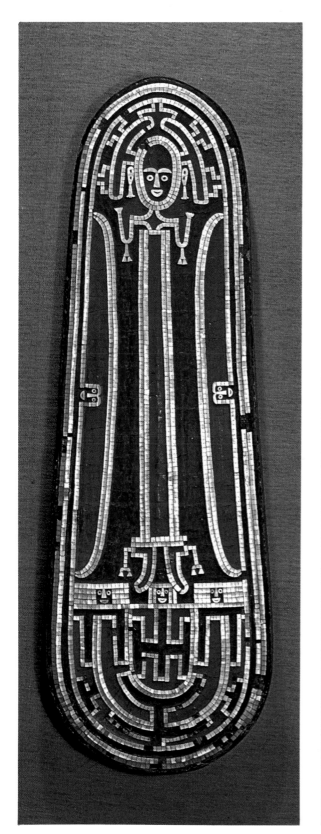

Weapons as Art

Looking at tribal art in general, and at weapons as art in particular, it is as well to keep in mind that there was no class of artists as such, or even artisans in most tribal societies. The commonly practised division of labour by sex generally meant that the production of weapons was an exclusively male art. Otherwise, everybody was his own craftsman (although some gifted individuals may have given away or traded their superior products). This accounts for the notably different levels of aesthetic quality among examples of an artifact within a given society. In some specific cases, however, the beginnings of professional specialization are apparent. Among some Plains Indian tribes, for example, visionary specialists would dream of shield designs and supply the best ones to the warriors. The production of metal weapons also tends to be more or less professionally specialized wherever they are used.

In our introductory discussion of tribal art it has already been remarked that seemingly useless decorations may serve a specific function for their maker or user. This applies particularly to weapons. Some designs reflect attempts to increase the warrior's success on a warpath or to protect him against adverse influences: the decorative patterns on shields discussed on page 85 are a prime example. Other decorations have developed out of the elaboration of functional attributes, as in many types of club. Still others have lost their original meaning with the passage of time, as traditional patterns are reinterpreted by successive generations.

A weapon that we find beautiful probably appealed to its maker because of its effectiveness. Tribal art is essentially concerned with the design and manufacture of artifacts: individual creativity is of minor importance. Most modes of artistic expression in tribal societies were circumscribed by the constraints of a traditional style, formed out of the cumulative experience of generations of makers and users of artifacts. Beauty was not sought for its own sake but was an integral part of an object's usefulness. On the whole, very little tribal art was not functional. Artifacts were made to fill certain needs of the individual or of society, and their shapes were primarily determined by practical considerations. What different people regarded as practical, however, might vary. A philosophy which took for granted a permanent relationship between man and the supernatural, for example, would find it eminently practical to allow for this in the design of an artifact. In this way, objects which do not appear to have a purpose may in fact have been functional to their makers. For a true comprehension of tribal art, therefore, knowledge of the cultural background is clearly mandatory.

104. Wooden shields from Queensland are among the most decorative works of art produced by the aboriginal Australians. Their brightly coloured non-representational designs are outlined in black and, in their remarkable diversity, represent the individual distinctions of warriors. *Berlin, Museum für Völkerkunde*

Bibliography

General Anthropological Studies of War
Bohannan, Paul, ed.: *Law and Warfare, Studies in the Anthropology of Conflict.* Garden City, New York, 1967.
Fried, Morton et al., eds.: *War: The Anthropology of Armed Conflict and Aggression.* Garden City, New York, 1968.
Frobenius, Leo: *Weltgeschichte des Krieges.* Hannover, 1903.
Nettleship, Martin H. et al, eds.: *War, Its Causes and Correlates.* The Hague-Chicago, 1975.
Turney-High, H. H.: *The Practice of Primitive War.* Missoula, 1942.

Ethnographic Studies of Tribal Warfare
Bell, F. L.: *Warfare among the Tanga.* Oceania 5:253–279, 1934.
Ferguson, R. F.: *The Zulus and the Spartans: A Comparison of their Military Systems.* Harvard African Studies 2:198–227.
Fernandes, Florestan: *A Função Social da Guerra na Sociedade Tupinambá.* 2nd ed., São Paulo, 1970.
Fortune, R. F.: *Arapesh Warfare.* American Anthropologist 41:22–41, 1939.
Gardner, Robert and Heider, Karl G.: *Gardens of War.* New York–Toronto, 1968.
Goodwin, Grenville: *Western Apache Raiding and Warfare.* Tucson, 1971.
Hill, W. W.: *Navaho Warfare.* Yale University Publications in Anthropology 5. New Haven, 1936.
Karsten, R.: *Blood Revenge, War, and Victory Feasts among the Jibaro Indians of Eastern Equador.* Bureau of American Ethnology, Bulletin, 79, 1923.
Knowles, N.: *The Torture of Captives by the Indians of Eastern North Amercia.* Proceedings of the American Philosophical Society 82(2), 1940.
Koch, Klaus-Friedrich: *War and Peace among the Jalémó.* Cambridge, MA, 1974.
Malinowski, Bronislaw: *War and Weapons among the Natives of the Trobriand Islands.* Man 20: no. 5, 1920.
Snyderman, George S.: *Behind the Tree of Peace. A Sociological Analysis of Iroquois Warfare.* Pennsylvania Archeologist 18(3–4), 1948.
Swadesh, Morris: *Motivations in Nootka Warfare.* Southwestern Journal of Anthropology 4:76–93, 1948.
Vayda, Andrew: *Maori Warfare.* Polynesian Society Maori Monographs 2, 1960.
Vayda, Andrew: *Phases in the Process of War and Peace among the Marings of New Guinea.* Oceania 42:1–24, 1971.
Warner, W. L.: *Murngin Warfare.* Oceania 1:457–477, 1931.
Wedgwood, Camilla H.: *Some Aspects of Warfare in Melanesia.* Oceania 1:5–33, 1930.

The Tribal Arsenal
Churchill, William: *Club Types of Nuclear Polynesia.* Carnegie Institution Publication 225, 1957.
Davidson, D. S.: *Australian Throwing Sticks, Throwing Clubs and Boomerangs.* American Anthropologist 38:76–100, 1936.
Dietschy, H.: *Die amerikanischen Keulen und Holzschwerter.* Internationales Archiv für Ethnographie 37:87–205, 1939.
Fischer, Werner and Zirngibl, Manfred A.: *Afrikanische Waffen.* Passau, 1978.
Krieger, Herbert W.: *The Collection of Primitive Weapons and Armor of the Philippine Islands in the U.S. National Museum.* U.S.N.M. Bulletin 137, 1926.
Lindblom, K. G.: *Fighting Bracelets and Kindred Weapons in Africa.* Smärre Meddelanden, Riksmuseets Etnografiska Avdelning 4, 1927.
Lindblom, K. G.: *The Sling, especially in Africa.* Smärre Meddelanden, Statens Etnografiska Museum 17, 1940.
Peterson, Harold L.: *American Indian Tomahawks.* Contributions from the Museum of the American Indian 19, 1965.
Pope, Saxton T.: *A Study of Bows and Arrows.* University of California Publications in American Archaeology and Ethnology 13:329–414, 1923.
Rex Gonzales, A.: *La Boleadora.* Revista del Museo de la Universidad Eva Perón 4:133–292, 1953.
Schmitz, Carl A.: *Technologie frühzeitlicher Waffen.* Basel, 1963.
Speiser, Fritz: *Über Keulenformen in Melanesien.* Zeitschrift für Ethnologie 64:74–105, 1932.
Stone, George Cameron: *A Glossary of the Construction, Decoration and Use of Arms and Armour in all Countries and in all Times.* (1934), reprinted New York, 1961.
Yde, Jens: *The Regional Distribution of South American Blowgun Types.* Journal de la Société des Américanistes, Paris, 37:275317, 1940.
Wistrand, C. G.: *African Axes.* Studia Ethnographica Upsaliensia 15, 1958.

Acknowledgements and list of illustrations

The author and John Calmann and Cooper Ltd would like to thank the museums and collectors who have allowed works from their collections to be reproduced in this book. They would also like to thank the photographers and photographic libraries who have allowed their photographs to be reproduced.

1. Lithograph of Dyak warrior. (Photo Fritz Mandl, Vienna)
2. Sioux warrior killing U.S. soldier, from *The Autobiography of Tall Bear.* London, British. Courtesy of the Trustees of the British Museum
3. Painted bison skin. Luzern, Collection Dr Hans V. Segesser-Epp
4. Australian aborigine attacking European. London, British Museum, (Dept. Natural History). Photo Axel Poignant
5. Bontoc Igorot Constabulary soldiers. Vienna, Museum für Völkerkunde (photo Philippine Bureau of Science)
6. A Turkoman attack on a Russian exploring expedition. (Photo Fritz Mandl, Vienna)
7. *Ball play of the Choctaw – Ball Down* by George Catlin. Washington, Courtesy of National Collection of Fine Arts, Smithsonian Institution
8. Xingu Indians wrestling. (Photo Heinrich Harrer)
9. Dani warriors fighting. (Photo Karl G. Heider)
10. Masai shield. Prague, Naprstek-Museum (photo Vera Jíslova, Prague)
11. Jalé boys learning to use bow and arrow. Photo Klaus-Friedrich Koch
12. An old man teaches grandson to use bow and arrow. (Photo Axel Poignant)
13. Ahaggar Tuarag fighting. Vienna, Museum für Völkerkunde (photo F. Chasseloup-Laubat)
14. Lithograph of Buru warriors duelling. (Photo Fritz Mandl, Vienna)
15. Lithograph of Bailundo warrior leaving for war. (Photo Fritz Mandl, Vienna)
16. Image of Kaka'ilimoku. Honolulu, Bernice. P. Bishop Museum
17. Zuni war god. London, British Museum. Courtesy of the Trustees of the British Museum
18. Hidatsa painted bison robe. Stuttgart, Linden Museum
19. Naga ancestor figure. Vienna, Museum für Völkerkunde (photo Fritz Mandl, Vienna)

Index